Energy Transition in East Asia

The Fukushima disaster of 2011 shook the globe, arousing warm debate and new research within the academic fields of countries in both the West and the East on issues related to nuclear security, public trust, government governance, risk governance and risk perception along with technological and social aspects. The Fukushima incident not only revealed the importance of risk governance in the East Asian region, but also became an important turning point in the restructuring of energy in several East Asian nations. However, the regulatory culture in East Asian countries is by nature different from that of their western counterparts; the history and culture of East Asia has formed East Asian countries' unique regulatory characteristics

This book aims to establish a risk governance structure for the East Asian region, providing a completely new perspective for both practical implementation and the academic field. It focuses on the problems of risk governance in East Asia. Through a discussion of the risk related issues raised by contemporary globalization, this book outlines the unique form of East Asia's risk governance architecture. It brings together the work of top academics from Japan, South Korea and Taiwan to provide a common picture of how these three countries' governments are dealing with the energy transition brought on by the climate change crisis. The various aspects of East Asia's unique regulatory culture and governance models are placed into context, while East Asia's risk governance theoretical framework is outlined.

Kuei-Tien Chou is Director of Graduate Institute of National Development and also Director of Risk Society and Policy Research Center, National Taiwan University. He got his doctoral degree in sociology from Universität München. His research interests are in the field of technological risk, climate change, risk governance, technological policy-making, public participation, science and democracy as well as East Asia modernity.

Routledge Contemporary Asia Series

For the full list of books: https://www.routledge.com/Routledge-Contemporary-Asia-Series/book-series/SE0794

Land Grabs in Asia
What role for the law?
Edited by Connie Carter and Andrew Harding

Asia Struggles with Democracy
Evidence from Indonesia, Korea and Thailand
Giovanna Maria Dora Dore

China-Malaysia Relations and Foreign Policy
Razak Abdullah

Politics of the 'Other' in India and China
Western concepts in non-Western contexts
Edited by Lion König and Bidisha Chaudhuri

China–India Relations
Cooperation and conflict
Edited by Kanti Bajpai, Huang Jing and Kishore Mahbubani

New Regional Geopolitics in the Indo-Pacific
Drivers, dynamics and consequences
Edited by Priya Chacko

Energy Security in Asia and Eurasia
Edited by Mike M. Mochizuki and Deepa M. Ollapally

Populist Threats and Democracy's Fate in Southeast Asia
Thailand, the Philippines, and Indonesia
William Case

Energy Transition in East Asia
A Social Science Perspective
Edited by Kuei-Tien Chou

Energy Transition in East Asia

A Social Science Perspective

Edited by
Kuei-Tien Chou

Routledge
Taylor & Francis Group
LONDON AND NEW YORK

First published 2018 by Routledge

2 Park Square, Milton Park, Abingdon, Oxfordshire OX14 4RN
52 Vanderbilt Avenue, New York, NY 10017

Routledge is an imprint of the Taylor & Francis Group, an informa business

First issued in paperback 2019

British Library Cataloguing in Publication Data
A catalogue record for this book is available from the British Library

Library of Congress Cataloging in Publication Data
A catalog record for this book has been requested

ISBN: 978-1-138-06574-1 (hbk)
ISBN: 978-0-367-37558-4 (pbk)

Typeset in Galliard
by Taylor & Francis Books

Contents

List of illustrations

Figures

Tables

Contributors

Koichi Hasegawa is Profesor of Department of Sociology, Graduate School of Arts and Letters, Tohoku University, Japan.

Dowan Ku is Director of Environment and Society Research Institute, Seoul, South Korea.

Kuei-Tien Chou is Director of Graduate Institute of National Development and also Director of Risk Society and Policy Research Center, National Taiwan University.

Sun-Jin Yun is Professor of Graduate School of Environmental Studies, Seoul National University.

Shu-Fen Kao is Assistant Professor, Department of Sociology, Fo Guang University.

Ying-Feng Chen is Assistant Professor, Department of Public Administration, Chinese Culture University.

Jin Hee Park is Professor, Dharma College, Dongguk University, Seoul, South Korea.

Miwao Matsumoto is Professor, Department of Sociology, Graduate School of Humanities and Sociology, The University of Tokyo, Japan.

Kohta Juraku is Associate Professor, Department of Humanities, Social and Health Sciences, School of Engineering, Tokyo Denki University.

Yuji Tateishi is Associate Professor, School of Sociology, Kwansei Gakuin University, Japan.

Mei-Fang Fan is Professor at the Institute of Science, Technology and Society, National Yang-Ming University; researcher at the Risk Society and Policy Research Center of National Taiwan University.

Chien-Ming Hsu is Assistant Research Fellow of Risk Society and Policy Research Center, College of Social Science, National Taiwan University.

Introduction: the cosmopolitan governance of energy transition

Kuei Tien Chou

We are suffering from global energy problems; climate change, nuclear power, peak oil, air pollution and so on are threatening the sustainability of global ecosystems and economic systems not to mention the survival of the human race itself. How can we survive such global energy risks? If we are ever to reconstruct an ecologically sustainable and socially just society worldwide, we must tackle the issue of energy transition. Energy transition can be defined as the structural change of energy production and consumption systems from an industrial paradigm with an energy system dependent on nature-exploitive energy sources to an ecological paradigm of an energy system making use of diversified, diffusive, and participatory energy sources. If the symbols of the industrial paradigm have been fossil fuel and nuclear energy, those of the ecological paradigm are solar and wind power.

Energy transition is not only an issue of a single country, but also regional and global issues; at the same time, it also involves several distinctive aspects, including the International Convention on climate change and carbon reduction, nuclear safety, moral choice, air pollution governance and sustainable renewable energy, all of which are trans-boundary, trans-discipline, and trans-regional, forming remarkable development globally. It has essentially become a part of globalized trans-boundary movements, posing great challenges to governance for all countries. These risk threats and challenges, on one hand, bring international pressure, vertically from top down, to societies. On the other hand, it provokes internal disputes and controversy over political choices, creating social pressures from bottom up. The diversified international political pressure and domestic social struggles interlink with one another and roll out endlessly. The eminent issues facing people are severe climate catastrophe, never-ending disputes over nuclear safety and nuclear waste storage, cross-border air pollution caused by fossil fuel combustion, the uncertainty of renewable energy prices on the international market, moral choice and social preferences and so on, forcing the globalized transboundary risk movement to develop in a growing spiral; energy transition, at the same time, takes this propulsion and pushes even further.

Why East Asia?

Global energy problems are provoked and strengthened by regional, national, and local levels of governance. East Asia is an important region in terms of global energy developments, challenging the industrial paradigm of the energy system. The societies of Japan, South Korea, and Taiwan are built, to a large extent, on the basis of fossil-based brown energy and produce high CO_2 emissions (IEA 2016). Similarly, in the history of these three countries, nuclear power accounted for a certain proportion of electricity generation. It is clear to say, the patterns of energy transition and the formation and varieties of social movement triggered by the Fukushima nuclear accident have significant meaning in the region.

The analytical method for this kind of globalization-transnational development requires a methodology that goes beyond the limits of a single country and puts the problems facing the three countries together. Methodological cosmopolitanism, as proposed by Beck and Sznaider (2006) and Beck and Grande (2010), suggests an approach of seeking a common governance model with different local characteristics rather than one confined to the individual, single political and economic structure within one national state. This enables analysis of the essential nature of energy transitions – transboundary features in the individual countries all around the globe.

While recognizing the importance of methodological cosmopolitanism in the search for the commonality of global problems, Beck argues that the analytical approach still should be initiated at the level of individual countries and go deep into the political and social complex of a country and its complicated political and economic structures (embedding the national). The goal is to facilitate the country's political transformation (replace the national) with the perspective of the cosmopolitan framework (Beck 2014a). This statement is rather significant, and it attempts to construct cosmopolitan governance by standing at a global level to examine the commonality and diversity of national problems. On this basis, we try to organize the authors from individual countries such as Japan, South Korea, and Taiwan, to analyse the character and structural problems of energy transition by embedding the issue in its wider political and social context. This is the first purpose of this book.

Furthermore, we attempt to extend the analysis framework, enriching the meaning of methodological cosmopolitanism from the angle of regional governance. That is to say, each region has its political, cultural, and geographical affinity, and even similar forms of technoscientific and economic development. In East Asia, Taiwan, Japan, and South Korea have a similar authoritative political history, the ensuing political democratization, and even their forms of technology, economy, and energy development are comparable. With this structure, they also produced a similar authoritative bureaucracy. Thus, in terms of cosmopolitan governance, we can further elaborate the regional political and social context (embedding the regional), and seek transformation within their common features and the structural challenges within the region (replacing the regional). This is the second and the main purpose of this book.

East Asian energy governance

Under this initiative, this book attempts to explore several pivotal questions in the region. First, authoritative bureaucracy or expert politics, shaped in the Cold War

context, constitutes a common structural problem in policy-making circles in these three countries. For example, the 'hidden risk society' demonstrated by the bureaucratic elite of scientific positivism in decision-making in Taiwan (chapter three), the 'structural disaster' incurred by bureaucracy in Japan (chapter seven), both of which have brought hidden and delayed risks, devaluating trust in government and general social trust. Secondly, in the field of energy, science, and technology, even in the face of the 311 Fukushima disaster, the decision-making structure of the politician-nuclear industry-scholar complex, which has long been developing under the above mentioned expert political structure, has seriously affected those countries' energy transition. In Japan, it is called 'nuclear village' (chapter one), while in South Korea, it is called 'nuclear mafia' (chapters two and four), and in Taiwan, it is called 'nuclear complex' (chapter three). Third, this bureaucratic elite complex in governmental decision-making in the East Asian countries, in addition to its entrenched dominant position in energy policies, has supported the ideology of developmental state or developmentalism since the 1970s. It promotes the ideology of economic development priority and pushes a brown economy with the massive use of brown energy; this even includes high-risk energy, such as the use of nuclear energy. On the one hand, with its tendency to a monopolistic or oligopolist national energy industry, it maintains the supply of low-cost electricity to the population so as to keep the low-income labour system alive; on the other hand, with the continuous propaganda framing sound nuclear safety, efficiency, and the low-cost myth, it produces the hidden, lagged nuclear accident risks in the long run. The situation hardly changed even when facing a civil society uprising over energy democracy resulting from the 311 disaster (chapters one, two, three, four, five, and six).

The factors highlighted above are situated in the case analysis of each country in this book, revealing a variety of similar policy culture of hidden risks and staying power, which is composed of institutional barriers and stagnant decision-making habits. It renders tremendous obstacles to energy transition. In other words, the locked-in-effects observed here must be analysed from the viewpoint of path dependency of social transformation. To explore these dilemmas found in the region's structural transformation and its possible breakthrough points, we need to emphasize the elements of social cognition, institutional arrangements, economic models, and administrative rationality as well as other political and economic complexes. The emphasis on the structural path-dependence phenomenon in this book forms the common scene of cosmopolitan governance which brings different case studies from each country, contributed by the authors of multiple nationalities, together.

Energy democracy in East Asia

The most prominent trend relating to cosmopolitan governance lies in the fact that after the Fukushima disaster, the social capacity of civil society in the three countries in East Asia was reignited, which is exemplified by the distinctive development of social robustness witnessed in each country. We can see that the appeal of energy transition gradually forms transboundary risk movement around the world, including the proactive choice of energy and fuel, the emphasis on CO_2 emissions, the struggle against coal-fired air pollution, the promotion of renewable energy, and the weight of social morality and value preference; all of these form the rapidly growing spiral of transboundary risk movement. In East Asia, this is reflected in the following three main shifts: the resurgence of the anti-nuclear campaign, the energy-saving movement, and the advocacy of renewable energy. More importantly, civil society and the combination of civil forces which promote energy transition from the bottom up have gained the ability to produce systematic socially robust knowledge. Through this thriving production of local and systematic risk knowledge, the civil society sector in the three countries is incrementally reinforcing their ability to challenge, lead, and even subvert the decision made by the bureaucratic elite and the current direction of energy policy.

Regardless of the divergent contexts of Taiwan, Japan, or South Korea, we can witness the growing and vigorous development of the energy democracy movement such as the 'One Less Nuclear Power Plant' movement, the grassroots anti-nuclear movement, civic-society-funded community power plants, issues of public health consideration, and finally the community-organized campaign against large-scale wind farms (chapters four, five, six, eight, and nine). In all the ongoing movements, in addition to the expansion and spread of energy citizenship and energy democracy, their characteristics of decentralization have challenged the existing dominant nuclear complex.

In particular, the analysis of the energy movements in these three countries has shown that citizens' groups have gained the capability to make transboundary knowledge discourse. Whether they are campaigns on anti-nuclear, concerning climate change and carbon emissions, radiation safety, windmill noise, energy-saving, public health, or about community power plants, local initiatives have gradually acquired the ability to put forward a discourse of knowledge and social morality, to compete with the discourse proposed by government or other large consortia. Jasanoff (2010) points out that on the issues of climate change, there is a need to re-focus from developing a large-scale understanding like the IPCC's climate knowledge to exploring its scale of meaning – the impact on human social action and knowledge, so as to build on the truly social meaning of climate disasters.

In fact, we can further argue that the issue of energy transition relating to climate change itself has the distinct feature of transboundary risk and social knowledge production. Similar to the concept of social robustness proposed by Nowotny (2003), and knowledge and moral competition suggested by Miller (2008, 2015), although both originated in western society, these two concepts have proved insightful to the analysis of the thriving progress of energy transition in the three East Asian countries. Finally, applying the perspective of embedding the regional to explore the regional features of cosmopolitan risk governance, we can also reveal the unique features of East Asian governance which is different from ones in the West; that is, the unique authoritative expert politics, hidden risk culture – the structure of 'nuclear complex', the reacting rising civil society, and the thriving social knowledge production and diffusion. These structural constraints and opportunities are incessantly forming important social logic and meaning, and they deserve continuous exploration.

Reference

Beck, U., and Sznaider, N. (2006). Unpacking cosmopolitanism for the social sciences: A research agenda. *British Journal of Sociology*, 57(1), 1–23.

Beck, U., and Grande, E. (2010). Varieties of second modernity: The cosmopolitan turn in social and political theory and research . *British Journal of Sociology*, 61(3), 409–443.

Beck, U. (2014). On methodological cosmopolitanism, Public lecture in the Seoul Conference2014 with Professor Ulrich Beck, Seoul.

IEA. (2016). *Key World Energy STATISTICS.* International Energy Agency.

Jasanoff, S. (2010). A new climate for society. *Theory Culture Society*, 27, 233–253.

Miller, C. A. (2008). Civic epistemologies: Constituting knowledge and order in political communities. *Sociology Compass*, 2(6), 1896–1919.

Miller, C. A. (2015). Knowledge and democracy: The epistemics of self-governance, In S. Hilgartner *et al.* (Eds.), *Science and democracy: Making knowledge and making power in the biosciences and beyond* (pp. 198–219). London: Routledge.

Nowotny, H. (2003). Democratising expertise and socially robust knowledge. *Science and Public Policy*, 30(3), 151–156.

Part I

Risk, energy and governance

Part I

Risk, energy and governance

1 Risk culture, risk framing, and nuclear energy dispute in Japan before and after the Fukushima nuclear accident

Koichi Hasegawa

The failure of risk culture and risk framing: a chain of underestimations, no defence against tsunamis and the station blackout

The Fukushima nuclear accident revealed the failure of Japan's risk culture and risk framing on nuclear energy.

The first immediate cause of the accident was an underestimation of the risk posed by tsunamis. The reactor buildings were constructed at an elevation of 10 metres above sea level after the site was dug down from an original elevation of 35 metres above sea level to secure solid foundations. The maximum tsunami height projected by Tokyo Electric Power Company (TEPCO) was 5.7 metres, but the actual inundation height at the time of the incident was 14–15 metres. Although there had been a number of warnings in previous years about the inadequacy of the assumed tsunami height of 5.7 metres, TEPCO, the Nuclear and Industrial Safety Agency (NISA) and the NSC neglected the risk and took no action (Soeda 2014).

The tsunami risk was not the only underestimation. A string of serious miscalculations had occurred, the most critical of which was the location of backup generators inside the oceanside turbine buildings, and above all, in the basement. It is incredible that the potential danger of installing backup generators in the basement of Japan's nuclear power stations with a high tsunami risk had not been considered a problem by any professionals for over 45 years until the Fukushima accident. The level of safety awareness among TEPCO, the NSC and NISA was so low that such an obvious question was ignored. The NSC did not have safety review standards concerning tsunamis until the occurrence of the Fukushima accident. There was virtually no defence against tsunamis.

The government and utility companies shared the assumption that there was no need to prepare for a station blackout, failing to provide emergency vent valves for a long period of time after construction (they were fitted to the reactors at Fukushima Daiichi Nuclear Power Station by 2001) and failing to anticipate the need for evacuation beyond a ten-kilometre radius.

None of Japan's electric power companies had anticipated a loss of all external power supplies and backup generators for a period of more than thirty minutes. In fact, the NSC gave this oversight a seal of approval without any evidence by stating in its inspection guidelines for the safe design of nuclear power plants formulated in 1990 that, 'There is no need to make allowances for a prolonged loss of all AC power supplies as power lines and emergency AC power generation facilities are expected to be restored to normal operation.' The NSC had not any evidence of reasoning 'expected to be restored to normal operation". The NSC was responsible for neglecting the need to prepare for the possibility of a station blackout.

The ten-kilometre evacuation zone

The NSC also made the mistake of underestimating the evacuation zone against the advice of the International Atomic Energy Agency (IAEA). Japan's guidelines for disaster prevention had postulated an evacuation zone of only eight to ten kilometres, despite a previous critical accident at Japan Nuclear Fuel Conversion Co. (JCO) in 1999. The guidelines were revised in May 2007 in light of the JCO accident and other considerations, but even this revision was perfunctory as Japan insisted that it had 'sufficient tolerance' against the advice of the IAEA for an urgent protective action planning zone of five to thirty kilometres. Every evacuation order failed to keep pace with the unfolding disaster and the evacuation zone was expanded many times as a stopgap measure because of this initial underestimation to a ten-kilometre radius. The then Prime Minister Kan issued instructions at 21:23 on 11 March that 'residents within a three-kilometre radius are ordered to evacuate and residents within a radius of three to ten kilometres are to stand by at home', on the assumption that the accident would be brought under control within twenty-four hours. He was forced to expand the ordered evacuation zone to a twenty-kilometre radius at 18:25 on 12 March.

The reasons of the third largest nuclear-powered country

As of the end of December 2010, Japan was the third largest nuclear-powered country in the world behind the US and France with a total of fifty-four reactors having a total installed capacity of 48.96 GW. At that time, nuclear power generation accounted for 30.8% of Japan's total generated electricity (the fiscal year 2010).

A fundamental reason for Japan's push toward nuclear power generation is the fact that it has the lowest energy self-sufficiency rate among the world's leading economies of only four per cent. As the Basic Energy Plan states, the government has promoted the use of nuclear power generation in order to achieve 'energy security' of stable supply of uranium, 'environmental suitability' of low CO_2 emissions and 'economic efficiency' with cheap costs to produce electricity (3Es).

Why has the only country to have experienced nuclear bombing been so intent on pushing forward with nuclear power generation? This is a question the author has frequently been asked by overseas scholars since the Fukushima accident. For example, Ulrich Beck, the leading proponent of the 'risk society' theory, asked, 'How could a country decide to develop nuclear energy without hesitation (...) while continuing to denounce the utter inhumanity of nuclear weapons tirelessly?' (Beck et al. 2011: 5–6).

The first answer is that the US targeted Japan as its prime market for the nuclear power generation, due to the very reason that it was the only country to have suffered nuclear bombing. The 'Atoms for Peace' speech made by the US President Eisenhower at the United Nations General Assembly in December 1953 was to serve two purposes: to sell licences and nuclear fuel to countries in the capitalist camp during the cold war and to prevent nuclear armament through the signing of nuclear energy agreements. Japan was the most effective stage from which the US could appeal to the world that the peaceful use of atomic energy was different from atomic weapons.

In Japan the words 'atom' or 'atomic' are mainly used in a civilian context such as 'atomic energy', 'atomic power station' and 'atomic industry' in a word-for-word translation. The usage of 'atomic bombing' is exceptional. The word 'nuclear' is mainly used in connection with armaments like 'nuclear weapons' and 'nuclear armament'. The usage of 'nuclear fuel' is exceptional. We never say 'nuclear power station' or 'nuclear energy' in Japanese.

Secondly, post-war Japan as a defeated nation and a resource-poor country harboured an excessive desire to rebuild itself based on science and technology. Japan sought to pursue prosperity through economic growth by relying on leading-edge nuclear technology. Psychological trauma carried by the defeated nation was behind the imperative to 'catch up' that characterized the social psyche of the time. Historical trauma played a role as well because Japan's need to secure oil resources had been the direct cause that triggered the Pacific War between 1941 and 1945.

The social mechanism of the proliferation of nuclear energy

What is the social mechanism that has supported the proliferation of nuclear power stations? Unlike the US, Japan has developed a shrewd system to promote nuclear power generation with the strong support of the central government as well as Korea and Taiwan. The number of nuclear reactors showed a steady increase in an almost straight line from the 1970s to 1997. Twenty units began operation in the 1970s, sixteen in the 1980s and fifteen in the 1990s. Nuclear plants were commissioned at a pace of almost two units per year until 1997. Japan's nuclear power plants had been constructed in a highly planned and 'socialistic' manner for a very long time. The process was called 'more socialistic than in socialist countries.'

The mechanism was similar to that for road construction. The Road Act system used to typify public works-related legal systems in Japan. A coherent system of legislation concerning planning, management, finance, quasi-government corporations and the environment was structured around the Road Act. A Five-year Plan for Road Construction had been formulated every five years since 1954. The plan carried qualitative and quantitative goals for road construction adopted by the Cabinet. Revising and updating the Five-year Plan for Road Construction was the focus of policy decision-making concerning road construction. Even though the construction of nuclear reactors involves private capital investment, it is in fact based on the government's 'socialistic' programme like road construction.

The counterparts of the Five-year Plan for Road Construction are the Long-Term Energy Supply and Demand Outlook and the Long-Range Programme on the Development and Utilization of Atomic Energy (renamed the Nuclear Energy Policy Guidelines from 2005). The former is deliberated and reported on by the Advisory Committee for Energy (renamed the Advisory Committee for Natural Resources and Energy from January 2001), established in 1965 as an advisory body to the Minister of International Trade and Industry (renamed the Minister of Economy, Trade and Industry from 2001), and serves as a de facto energy policy rather than a mere supply and demand outlook. The outlook has been updated every three to four years containing a programme for the next fifteen to twenty years and provides a basis for the siting of power stations. The fact that policy decisions are made by the Cabinet without Diet deliberations demonstrates its virtual policy status as with the Five-year Plan for Road Construction. There is a high level of coherence between the stable energy supply policy of the public utility departments of the Agency for Natural Resources and Energy at the Ministry of Economy, Trade and Industry (METI; former MITI) and the electric power supply policy of each utility company.

Ineffectiveness of safety regulations

Japan's nuclear energy administration had a structural problem in that the Nuclear Safety Commission (NSC) and the Nuclear and Industrial Safety Agency (NISA) did not have enough independence to effectively administer safety regulations. The IAEA recommended prior to the Fukushima accident that the NSC and NISA should be more independent, but the government ignored their recommendation. It was only after the Fukushima accident that the government began to seriously consider making the NSC more independent and separating the NISA from the METI. Finally, on 19 September 2012, the NISA and the Nuclear Safety Commission were merged to become the Nuclear Regulation Authority (NRA) and its secretariat called the Nuclear Regulation Agency as an affiliated organization of the Ministry of Environment, and the NRA is now separated from the METI on a superficial level. But most of the staff of the NRA was provided by the METI and former NISA.

Three investigative reports have been published on the Fukushima accident: one was produced by the national governmental committee, the Investigation Committee on the Accidents at Fukushima Nuclear Power Stations of Tokyo Electric Power Company (ICANPS 2012); the second by the Diet, the National Diet of Japan Fukushima Nuclear Accident Independent Investigation Commission (NAIIC 2012); and the third by the private investigatory committee, the Independent Investigation Commission on the Fukushima Nuclear Accident (2014).

The NAIIC report (2012: Ch. 5, 47 and 48) focused mainly on regulatory and other organizational factors in the case of the Fukushima accident: '[The Japanese regulators] stuck to their belief of infallibility so much that they were reluctant to improve safety regulations, and thus their mindset was structurally ill-matched for running a safety culture.' This 'safety culture' existed in name only, and as a result, the notion of safety and security was 'sold', cheaply and irresponsibly, to the whole nation.

This report also criticized the regulatory agencies' lack of independence, transparency, and expertise, by using the term 'regulatory capture', meaning 'regulatory authorities become the '"captives"' of nuclear operators as they are devoted to maximizing the interests of the nuclear operators that are supposedly placed under them (NAIIC, 2012: 12).

Rigidity and self-preservation of the 'nuclear village'

Nuclear power generation stands out in terms of the delaying of problem resolution due to stonewalling by vested interests, the use of stopgap measures and the lack of risk awareness among policy-makers. Bureaucracy in general has a self-preserving nature, but the nuclear industry, like the arms industry, is evidently innately self-preserving. The nuclear industry has a similar character to the arms industry as it has its origins in the civilian application of military technology and it cannot easily be converted to another type of business. Constructing nuclear power stations is necessary for the survival of the nuclear industry itself.

In the case of Japan in particular which, unlike the US and France, does not conduct nuclear development for military purposes, it is essential to maintain programmes for nuclear power plant construction and the nuclear fuel cycle so that the nuclear energy division of major corporations such as Mitsubishi Heavy Industries, Hitachi and Toshiba can survive.

Organizations that do not adopt a top-down system of management such as the bureaucracy in Japan are generally reluctant regarding any policy change as top executives do not have the freedom to assert leadership and significant time is spent on internal coordination and consensus formation. Especially under the job rotation system of Japanese ministries and agencies where fast-track bureaucrats are moved to new disciplines every few years, they are expected to follow existing policies and procedures for the period of their tenure and rarely take risks for fear of jeopardizing their promotion. When a majority of the top and middle managers behave this way, the whole organization suffers from an 'accumulated ostrich policy' as a result (Hasegawa 2015: 25).

It is particularly difficult for insiders to instigate reviews of politically important questions that are highly contested by external parties, such as the pros and cons of nuclear power generation. Unless there is what I call 'overriding social pressure' (Hasegawa 2015: 25), which is some form of social pressure to take urgent action that comes from outside an organization, including international pressure or proceedings, judicial decisions, disasters, accidents and intervention by powerful ruling party politicians, it is very rare that policy change is proposed or consensus is reached within an organization.

Long-standing closed relationships between politicians, bureaucrats, nuclear industry executives and the mass media – referred to as the 'nuclear village' (Hasegawa 2015: 22) – had received much criticism even before the Fukushima accident. In Japanese society, 'village' refers to traditional paternalistic relationships characterized by secrecy and exclusive vested interests. Notably, 'village' opposes the concept of civil society, which is autonomous, horizontal and open to all the public.

After the accident, the media published a self-reflective study, providing a detailed historical investigation into the media's role in these relationships, and their own contributions to encouraging the nuclear 'safety myth' (Jomaru 2012).

The secret of the nuclear budget

The continuation of the rigid policy process is supported by the underlying fiscal mechanism and the system of closely linked vested interests. Large public works projects are unstoppable thanks to the entrenchment of self-preserving systems backed by revenue sources and vested interests.

The revenue source for roads is another typical example. The 'specified revenue source' earmarked for road construction was the source of funding for road building under such laws as the Gasoline Tax Law and the Local Road Tax Law (the Local Gasoline Tax Law from 2009). (Although the specified revenue source for road systems was abolished in 2008, the situation has not changed much.) For instance, when a litre of gasoline is purchased, a gasoline tax of 53.8 yen (48.6 yen earmarked for national road construction plus a local tax of 5.2 yen for prefectural and municipal road construction) is included in the price. A majority of Japanese citizens are unaware of this mechanism that ensures that more roads are constructed alongside increased usage.

The nuclear energy-related annual budget of the METI and the Ministry of Education, Culture, Sports, Science and Technology (MEXT) combined is 432.3 billion yen (2010 budget bill). For electricity, there is an Electric Power Development Promotion Tax of 0.375 yen per kWh, just under two per cent of the domestic electricity unit price, which is automatically included in power bills and finances the Special Account for Energy/Electric Power Development Promotion (formerly the Special Account for Electric Power Development Promotion). In the cases of both roads and power plants, the government automatically secures revenue by way of tax-inclusive pricing without informing consumers, as the electric power development promotion tax is not even shown on the detailed power bill and therefore a majority of consumers are not aware that they are paying it. It amounts to 1,350 yen per year for an average household using 3,600 kWh of electricity per year. The good-natured Japanese consumers remain 'uninformed' about the fact that they have been paying this as part of their electricity charges since the introduction of this system in 1974 and thus promoting the construction of nuclear power plants. The total revenue in the Special Account for Energy/Electric Power Development Promotion is about 316.2 billion yen (2.84 billion US dollars), of which about 179 billion yen (1.61 billion dollars) is allocated to the so-called '*dengen sanpō kōfukin*' (grants by the three laws for power facility development) to cover the costs of siting power plants (2010 budget bill).

The grants by the three laws for power facility development is a system set up by the MITI in 1974 right after the first oil shock during Kakuei Tanaka's cabinet and constitutes the greatest weapon for the siting of power plants. The grants are administered under the Electric Power Development Promotion Act, the Act on Special Accounts for Electric Power Development Promotion Measures (presently

the Act on Special Accounts) and the Act on the Development of Areas Adjacent to Electric Power Generating Facilities. This mechanism of distributing 'compensation for inconveniences' was established when the siting of nuclear power plants began to encounter strong resistance and opposition from residents who were concerned about safety alongside the growing awareness about pollution and environmental issues from the mid-1960s. The grants are handed out to the district hosting the power plant and its surrounding municipalities from the start of construction and for five years after commissioning in the case of conventional power plants and ten years or so in the case of nuclear power plants and are used to build community facilities such as cultural, educational and sports facilities and roads. The system was consolidated in 2003 as the subsidy for power plant siting, which could be administered in a more flexible manner. The subsidies continue to be distributed for a wide range of purposes from the year after a siting feasibility study is undertaken prior to the start of construction until decommissioning. When the siting of a 1.35 GW nuclear power plant is accepted, the host and adjacent municipalities, including the prefectural government, cities, towns and villages, receive a windfall of 121.5 billion yen (1.1 billion dollars) in total over a period of forty-five years or so from the year after the feasibility study to the end of operations (Shimizu 2011a: 677).

The US and Europe do not have a grant/subsidy system such as this. Modelled on the Japanese system, Taiwan established the Act on the Administration of the Fund for Electric Power Development Promotion in 1988 and Korea enacted the Act on Support for Areas Neighbouring Power Plants in 1989 in order to have their electricity companies contribute part of electricity charges to deal with local communities (Shimizu 2011b: 102).

Courts in favour of the government

There have been numerous administrative lawsuits and civil actions to stop the construction and operation of many nuclear energy facilities such as nuclear power plants, fast-breeder reactors and nuclear fuel cycle facilities. However, Japan's courts are notably inclined to maintain the status quo and decide in favour of the defendant in lawsuits concerning nuclear facilities. Out of about twenty cases over a period of more than forty years, the plaintiff won in only two cases prior to the Fukushima accident: the Monju ruling handed down by the Nagoya High Court's Kanazawa branch in January 2003 and the Shika Unit No. 2 reactor decision by the Kanazawa District Court in March 2006. Yet, higher courts subsequently overturned both decisions and the plaintiff lost. There has been no instance of even temporary suspension of construction work by these actions (see below).

This contrasts sharply with the situation in Europe and the US in which court rulings are playing a major role in the closure of nuclear power plants and nuclear facilities.

Weak social watchdogs

Although nuclear energy administration was split between the Science and Technology Agency and the MITI before the reorganization of ministries and agencies, the decision-making process for planning commercial nuclear power plants was very much unified and the coherence of the overall nuclear energy policy improved further after the reorganization in 2001. The Diet had very limited opportunities to get involved in either the system or the budget and interest in energy and nuclear power issues was generally low among politicians of both the ruling and opposition parties until the Fukushima accident, aside from those representing constituencies with nuclear power plants or the electricity and associated labour unions.

The Liberal Democratic Party (LDP) has been a long-time promoter of nuclear power generation and receives large political donations from the electric utility companies. The Democratic Party of Japan (DPJ) has been rather equivocal in its attitude towards nuclear energy. Many of its members are backed by the Federation of Electric Power Related Industry Workers' Unions of Japan or the Japanese Electrical Electronic and Information Union consisting of labour unions at electric machinery companies such as Toshiba, Hitachi, Mitsubishi and Panasonic, and they have been strong supporters of nuclear power generation.

The Social Democratic Party (SDP) has been opposed to nuclear power generation and the Japanese Communist Party (JCP) also has a negative view, but the issue has never been a major point of contention in the Diet partly because of the long-standing single party dominance of the LDP.

In the US, nuclear power generation has been one of the typical battle lines dividing the policy stances of the supportive Republican Party and the opposing Democratic Party for a long time along with economic policy and abortion. Due to close rivalries between the two parties in the presidential, upper and lower house, state governor and state legislature elections, the nuclear power generation issue has been the focus of political confrontation until recent years.

In Japan's nuclear energy administration, nuclear-related laws form a separate legal structure headed by the Atomic Energy Basic Law (promulgated in 1955), which is characterized by coherence and 'untouchability' as it has been outside of the regulations under the former Basic Act for Environmental Pollution Control and the current Basic Environment Law. The only involvement the former Environment Agency had until the reorganization was in terms of the regulation of thermal discharges. After the reorganization, the Ministry of the Environment assumed jurisdiction over the monitoring and measurement of environmental radioactive substances and proper treatment of radioactive waste. It was only in Japan as well as Taiwan and Korea, which were modelled on the system in Japan, that a central ministry or agency in charge of environmental policy had little regulatory power over radioactive material or waste.

Japan was the last country to pass the Environmental Impact Assessment Law among twenty-nine OECD member countries (at the time) in 1997, some twenty-one years after the first bill was submitted in 1976. The difficult passage of this law was caused by persistent resistance from the electricity industry and the MITI, which feared that the legislation would slow the process of power plant construction. The final version of the law treats power generation as a special case and gives substantial concessions to the initiatives of the MITI, including ministerial recommendations and amendment orders.

After all, the Japanese system of 'nuclear energy promotion' provided very few opportunities for outside parties to carry out independent checks on nuclear power stations. Because of the weakness of all types of social watchdogs, including consumer groups and environmentalist organizations, nuclear power stations have been treated as sacred cows and open criticism has been regarded as a kind of taboo. The mass media has been deferential to the electric utility industry as it has been one of its biggest sponsors. The level of efficacy of social monitoring over nuclear power generation is a barometer of democracy in a society and one of the greatest challenges for post-Fukushima Japan.

Risk culture and risk framing after the Fukushima accident

What changes has the Fukushima accident caused? More than 80,000 people lost their homes, their beautiful farmland on which many generations of their ancestors had worked ardently, their fisheries, their customers, their workplaces and their hope. Their communities were divided and many had their family lives destroyed (Hasegawa 2012). What has Japanese society gained in return for these losses? What changes are we observing? Japan's risk culture and risk framing are examples of notable changes. Nuclear energy dispute has also changed.

The author shows good examples how risk culture and risk framing have been changed after the accident.

Figure 1.1 reveals the changing opinion on nuclear power generation between 1978 and 2015. 1) Before the Fukushima accident, more than half supported increasing the number of nuclear reactors. Only 16–17% supported 'decreasing' in 2005 and 2009. 2) Right after the accident, supporters of 'increasing' drastically decreased to only a few per cent. 3) June 2011 was another turning point in public opinion. Then more than 60% supported 'decreasing' the number of nuclear reactors. On 6 May 2011 the then Prime Minister Kan requested that Chūbu EPCO stopped the operation of all reactors at the Hamaoka Nuclear Power Station located in the epicentral area of an anticipated Tōkai earthquake. In response, the company accepted this request. In May and June 2011 people started to realize the severity and the long-term effects of the accident. 4) Since July 2011, around 70% have been supporting 'decreasing'. This pattern of opinion has been very stable in polls of any media.

Professor Hirotada Hirose conducted questionnaire surveys on reliable and unreliable information sources in September 2010, June 2011, March 2012, March 2013, and August 2013 as shown in Figures 1.2 and 1.3 (Hirose 2013). The answers before and after the accident show a sharp contrast. (1) Before the accident, 19% of citizens felt mostly the central government was 'reliable' and 30% felt local governments were 'reliable'. But after the accident, each ratio dropped by 10%. (2) Reliability of the central government recovered in August 2013. (3) After the accident, reliability of international institutes and professionals in universities and research institutes increased. (4) After the accident, more than half of citizens felt central government sources were mostly 'unreliable'. This ratio went up more than 30 %, from 19.7% before the accident. Before the accident, the ratio of

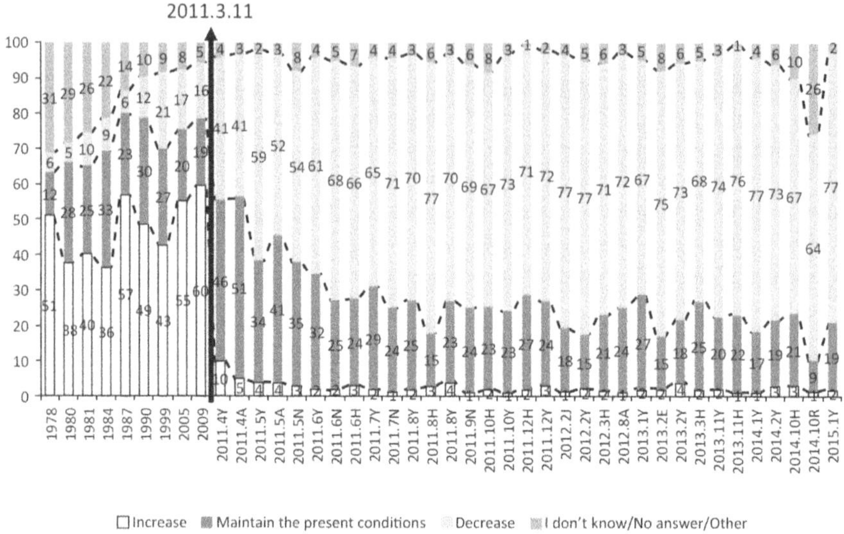

Figure 1.1 Opinion on nuclear power generation 1978 to 2015
Source: Iwai and Shishido (2015: 183).

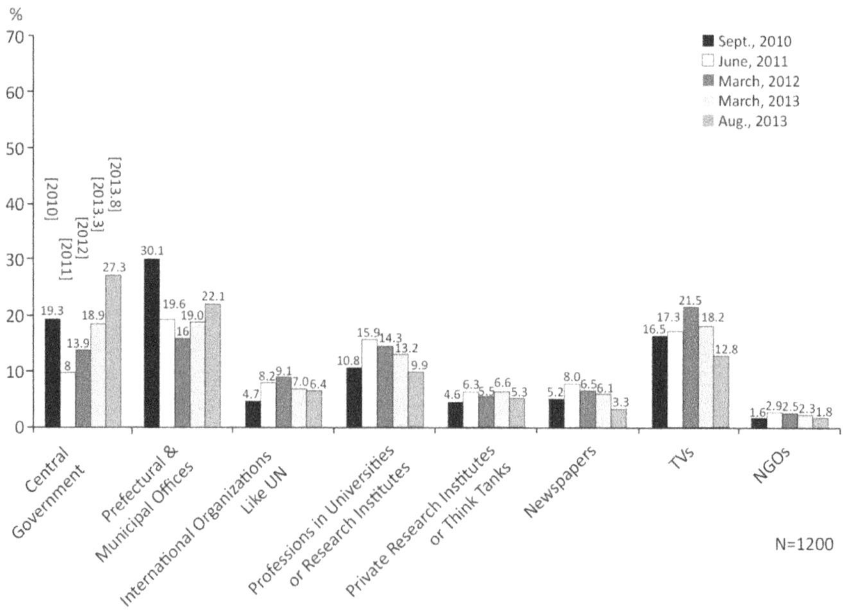

Figure 1.2 The most reliable information source
Source: Hirose (2013: 1347).

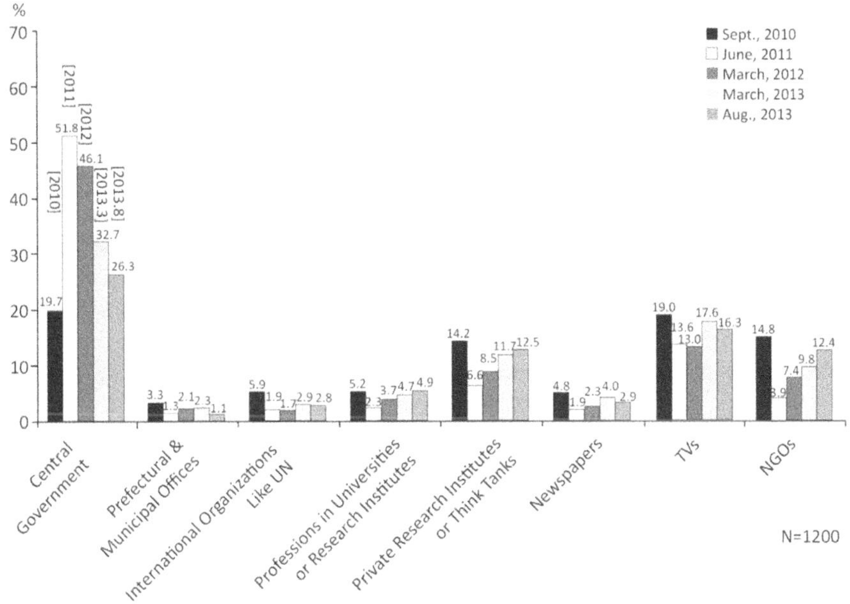

Figure 1.3 The most unreliable information source
Source: Hirose (2013: 1347).

'reliable' and 'unreliable' to the central government was balanced. But three months after the accident, we can observe the sharp contrast of 'unreliable' 52% vs. 'reliable' 10%. (5) Before the accident, the ratio of 'unreliable' to private think tanks, TV, and NGOs were relatively high. People were sceptical about the 'private' institutions. But after the accident the ratio of 'unreliable' decreased more than 10% to NGOs, 8% to private think tanks and 5% to TV, although the ratio of 'unreliable' returned to the almost previous level in August 2013.

Upheavals of citizens' protests

After the accident, energetic citizens' protests with several tens of thousands of protesters became frequent in 2011 and 2012 motivated by strong distrust of information disclosure by central government and the power company TEPCO, strong anxiety about the effects of radiation, and great anger over the Prime Minister's disregard of anti-nuclear sentiments (Hasegawa 2014).

In Japan, for more than 50 years, the largest anti-government activity was the anti-Japan US security treaty campaign in June 1960 with 330,000 citizens involved (organizer's announcement). After the defeat of this campaign, occasions where more than 10,000 protesters gathered were very few for any agenda. In case of the nuclear energy issues, in April 1988, 20,000 protesters gathered in Hibiya Park, Tokyo. This was the largest anti-nuclear protest. On 9 April 1989, in Rokkasho village, Aomori, 10,000 protesters gathered against the construction of nuclear cycle facilities including a reprocessing factory where plutonium is extracted from spent nuclear fuel.

Civil society in Japan is changing into one where anyone can easily participate in demonstrations. The incidents of the evidence followed. In July 2014, the Abe Cabinet approved the use of the right to collective self-defence and the Diet passed a series of related acts in September 2015 although opinion polls show the majority is opposed to these and several tens of thousands of citizens protested against these acts outside the Diet, and in many local cities nationwide. These series of protests were strongly affected by anti-Fukushima protests in 2011 and 2012, regarding styles of organizing protest activities.

A series of protest activities in Japan following the accident had some similarities to the protests in 2011 that spread from the Arab world to Europe, and the United States, showing their contemporaneity. Some of these similarities included the use of smartphones or tablet computers, and social networking services like Twitter, Facebook, YouTube, and Ustream to mobilize people (Lotan et al. 2011). Small groups and individuals were mobilized on a voluntary basis, without relying on existing political parties or labour unions. Another common characteristic was encouraging protesters to express their individual feelings, thoughts, and ideas regarding protest activities, rather than directing them to work toward the achievement of concrete and clearly defined goals.

The following facts on protest activities are remarkable regarding the history of civil society and social movements in Japan: (1) protests are based on people's voluntary participation, not on organized mobilization by major labour unions or other organizations; (2) large-scale street protests have taken place in the areas surrounding the Diet building, where strict regulations prohibit activities like the use of megaphones; (3) the so-called Kantei demonstrations which were held in front of the Prime Minister's office have been peaceful, and there were only two arrests; (4) the weekly Kantei demonstrations have continued for more than five years; (5) protests have spread across Tokyo to other cities nationwide; and (6) civil society in Japan is changing so dramatically that anyone can participate in a demonstration against government policy (see Hasegawa 2014).

Many court actions on nuclear power have been fought in various parts of Japan over the years thanks to the efforts of local residents' movements around nuclear facilities and lawyers. But as the author described earlier, most of the twenty or so cases in total have ended in defeat for the plaintiffs. There were only two successful cases prior to the Fukushima accident. Both cases were subsequently defeated in higher courts.

After the Fukushima accident, a lawsuit was initiated in November 2012 seeking an injunction on the operation of Ōi Units No. 3 and 4. The Fukui District Court handed down an order to stop the operation on 21 May 2014. It was a historic ruling that earnestly addressed the Fukushima accident. On 14 April 2015, the Fukui District Court issued a provisional injunction ordering Kansai EPCO not to restart Units No. 3 and 4 at the Takahama Nuclear Power Station in Fukui Prefecture. With the injunction taking effect immediately, the utility is prohibited from restarting the reactors unless the injunction is reversed in future appellate procedures. This is the first and an historical court-ordered injunction disallowing the operation of reactors.

These upheavals of citizens' protests and the legal decisions above sharply show how risk culture and risk framing have changed following the accident.

Nuclear energy dispute after the accident

All nuclear units in Japan were closed temporarily for 23 months from 15 September 2013 to 11 August 2015. There were no troubles nor any power outage and shortages. Especially within TEPCO and Tohoku EPCO areas, there was no nuclear operation following the 3.11 disaster. As of the end of January 2017, only three units are operating under the approval of the new regulatory agency, the Nuclear Regulation Authority (NRA). Due to the legal requirement for regular maintenance, a nuclear unit has to close temporarily after operating for at most 13 months. Before 15 September 2013, many units had to close temporarily on a one by one basis for regular maintenance. Some units along the Pacific Ocean were forced to close because of direct damage sustained from the earthquake. The NRA was established on 18 September 2012, replacing the Nuclear Safety Commission and the Nuclear and Industrial Safety Agency, both of which had failed to prevent the accident, and revealed their mal-functioning through the Fukushima accident. The NRA is expected to be independent of the national government.

After the accident, the focal point of the nuclear energy dispute shifted. The potential for electricity shortage without nuclear power in the short term became less important. The Thermal reactors replaced the supply of electricity formerly supplied by nuclear generation. The price of electricity has increased by several per cent, but the majority of people support decreasing the number of nuclear plants as shown in Figure 1.3.

We have three options for nuclear energy policy; (1) The immediate closure of all nuclear reactors, supported by former Prime Minister Koizumi. (2) Reopening the nuclear reactors which can meet the new standard set by the new regulatory agency, admitting but strictly limiting the 40-year operation of each reactor and rejecting the construction of new reactors, so aiming for denuclearization by the end of the 2030s. The former DPJ cabinet set this policy but the current Abe cabinet abandoned this policy. (3) Reopening the nuclear reactors which can meet the NRA's new standard and aiming to construct new reactors and replace old reactors. The Abe cabinet supports this policy and wants to keep the level of electricity produced by nuclear at around 20–22% of the total electricity supply by 2030.

Overestimation by METI

In July 2015, METI's long-term outlook said the total power generation of 1065 billion kWh in the fiscal year 2030 will be provided by 22–24% of renewable energy, 20–22% of nuclear, 27% of LNG-fired thermal, 26% of coal-fired and 3% of oil-fired. The basic assumption is that economic growth will be continuing to average a 1.7% increase per year from Fiscal Year 2013 to 2030. The required power generation will be 1278 billion kWh in 2030 and it will reduced by 17% to 1065 billion kWh through power conservation (see Figure 1.4). What is wrong?

Figure 1.5 shows the reality, total power generation peaked at 1033 billion kWh in 2007 and it declined by 14.1% and reached 885 billion kWh in 2015, equivalent to the level in 1995. 1278 billion kWh in 2030 is a complete overestimate, a 36% increase from 2013 in spite of a depopulated society and a low economic growth averaging a 0.7% increase of GDP per year between 2010 and 2015. The projection of 20–22% of energy generated by nuclear, 213–234 billion kWh in financial year 2030 can be replaced by power conservation with no carbon emissions.

Since the Fukushima accident citizens have realized that nuclear energy is unnecessary in both the short and long term. Other main reasons for opposing nuclear energy in Japan are (1) a high risk of accident and disaster especially in a quake-prone archipelago, (2) financial risk, (3) risk of handling spent nuclear fuel, and (4) risk of nuclear armament.

Risk of nuclear reactors in a quake-prone archipelago

The Fukushima accident has posed the most serious problem that a major 'quake-prone country' and nuclear power generation are incompatible. Japan is a major earthquake-prone country due to its location on the boundaries of four tectonic plates. As Katsuhiko Ishibashi (2011) points out, fifty-four reactors or 12.2% of the world's commercial nuclear reactors are situated in this small islands nation

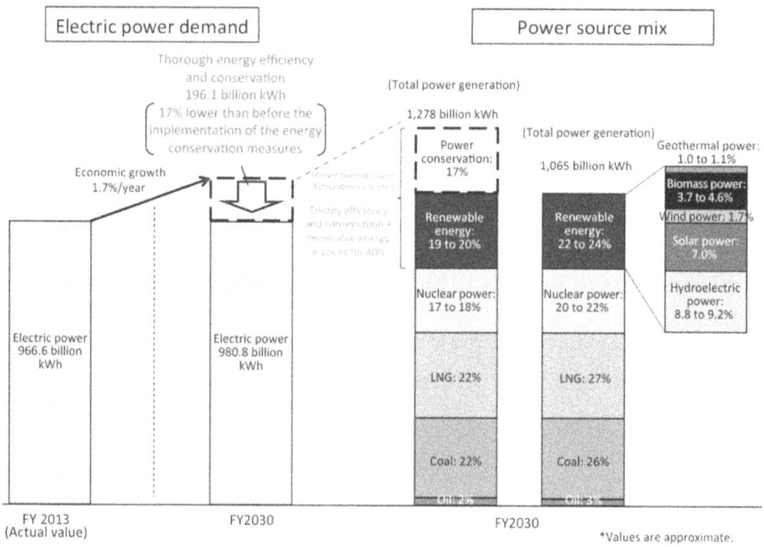

Figure 1.4 Long-term energy supply and demand outlook
Source: Website of the Ministry of Economy, Trade and Industry of Japan (http://www.meti.go.jp/english/press/2015/0716_01.html).

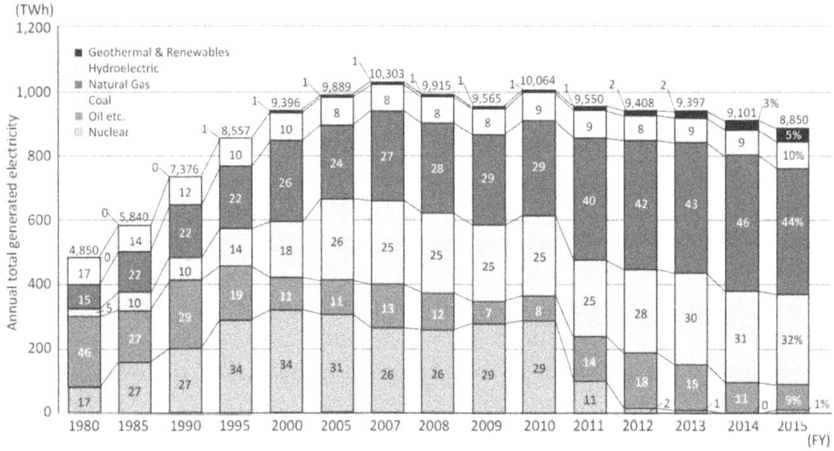

Figure 1.5 Total power generation 1980 to 2015

Source: 'Historical Trend of Power Generation Volume by Source in Japan' in *Graphical Flip-chart of Nuclear & Energy Related Topics.* (http://www.ene100.jp/www/wp-content/uploads/zumen/e1-2-7.jpg) Authorized by Japan Atomic Energy Relations Organization (JAERO).

where about 10% of the world's seismic activity occurs, and moreover, the area has entered a seismically highly active phase. He also advises that the Great East Japan Earthquake has raised the probability of major earthquakes in almost all parts of the Japanese Archipelago, claiming 'It will certainly be "a combined earthquake and nuclear disaster"' (Ishibashi 2011).

Financial risk for utility companies and plant makers

'Nuclear energy is safe, clean and cheap.' The Fukushima accident revealed this propaganda to be an illusion. It brought significant additional costs to cover the new requirements for improving safety. Although TEPCO is the largest private utility company in the world, since the Fukushima accident it has faced bankruptcy by being obliged to pay a total of 8,000 billion yen (72 billion dollars), annually 300 billion yen (2.7 billion dollars) as the fee for decommissioning and total of 3,900 billion yen (35.1 billion dollars), annually 200 billion yen (1.8 billion dollars) as compensation to the refugees and sufferers. After the accident TEPCO has been kept afloat by the Japanese government. If a severe nuclear accident happens again, the power company will be liquidated

Toshiba, another leading giant company, is also facing a potential crisis of bankruptcy as of the end of January 2017 due to the manipulation of its accounts of 230 billion yen (2.1 billion dollars), the loss of 500 billion Yen (4.5 billion Dollars) in the 2015 FY and an additional new loss of 700 billion yen (6.3 billion dollars) resulting from its failure to buy another US nuclear manufacturer in FY 2016. The initial trigger was buying the nuclear energy division of Westinghouse in 2006. At that point Toshiba became the world's largest nuclear reactor manufacturer. But it was misleading and the beginning of the tragedy. Westinghouse used to be a leading general electric appliances manufacturer in the US and was considered a major rival to GE. It developed the pressurized-water reactors (PWR) that were used by KEPCO and others in Japan. Westinghouse gradually declined and became defunct in 1999, but it sold its nuclear energy division to BNFL in 1998. BNFL was driven to the brink of bankruptcy and its owner, the UK government, decided to let Westinghouse go. Toshiba won the bid and paid five billion dollars, double the estimated value, in anticipation of a nuclear renaissance. In 2009 Toshiba declared its goal of getting orders for 39 new reactors globally by 2015. In reality, Toshiba started construction of a total of only eight reactors in China and the US from 2006 to 2015. The Fukushima accident dealt the final blow to the nuclear renaissance, which was already under dark clouds by 2010.

Handling spent nuclear fuel and three endemic problems

The greatest dilemma facing every country is how to dispose of spent nuclear fuel. The impossibility of resolving spent fuel disposal prompted a move away from nuclear energy in the US and Europe. Japan has adopted a closed nuclear fuel cycle policy by which all spent fuel is reprocessed and high-level radio-active waste is stored for cooling over a period of thirty to fifty years in a vitrified form and eventually disposed of by deep burial in a geological formation 300 to 700 metres below ground. Handling high-level radioactive waste is extremely cumbersome and dangerous due to extremely high levels of radioactivity, large amounts of heat generation, high toxicity, very long half-lives, and the presence of a wide variety of elements. It needs to be isolated from the human living environment for at least 100,000 years or so and hence requires permanent storage facilities. But every country agonizes over the location of final disposal sites.

Japan's difficulties on spent nuclear fuel are compounded by three other endemic problems. First, it is questionable as to whether there are any suitable sites for final disposal in an earthquake-prone country which we realized on 11 March 2011 with interlaced active fault lines, a limited land area, high population density and many subterranean water veins.

The second problem is the issue of surplus plutonium. Japan has the highest potential capability for nuclear armament in the world and the IAEA has been most concerned about the possible use of surplus plutonium to produce atomic weapons. For this reason, in 1991 the Japanese government announced its commitment to not hold surplus plutonium. In order for Japan to adhere to its two principles, namely, the full reprocessing of all spent nuclear fuel and zero surplus plutonium, it needs to strictly maintain the demand and supply balance for plutonium extracted by reprocessing. The dilemma is that reprocessing needs to be stopped in order to prevent the surplus plutonium issue, but this cessation will breach the principle of full reprocessing of all spent nuclear fuel. The Fukushima accident makes this dilemma more problematic (see Hasegawa 2015: 51–53).

The third problem is the difficulty of changing policy despite the high cost of reprocessing. Why is Japan so obstinately fixated on reprocessing? The ulterior motive is to retain international rights relevant to the potential military use of nuclear resources in the future. As Germany is phasing out its nuclear programme, Japan is the only non-nuclear-weapon state that is allowed to possess technologies for uranium enrichment, reprocessing and fast breeder reactors. The fundamental premise of the nuclear fuel cycle is an expansion of nuclear power generation. The Fukushima accident is forcing Japanese society to face the challenge of a radical rethinking of its nuclear fuel cycle programme.

Potential nuclear deterrent and risk of nuclear armament

Another reason Japan continues down the path of nuclear power generation is that having nuclear power, especially nuclear fuel cycle capability, is considered to be a potential nuclear deterrent.

The Japan–US Nuclear Cooperation Agreement that was revised in 1988 gave Japan 'blanket consent' with regard to spent nuclear fuel reprocessing to process any amount up to a certain limit as it wished. The current agreement will expire in July 2018. How significant is it for Japan to have the potential to arm itself with nuclear weapons? Japan's nuclear cooperation agreement with the US, Canada and Australia limits the use of nuclear fuel to that for 'peaceful purposes'. The official position adopted by Japan's past cabinets has been that 'possession [of nuclear weapons] does not violate the Constitution if it is limited to … a bare minimum required for self-defence.' Legal restrictions on the possession of nuclear weapons domestically include Article Two of the Atomic Energy Basic Law, which states that it is '… limited to peaceful purposes', and the so-called three non-nuclear principles adopted by the Diet. In terms of international law, Japan is bound by nuclear cooperation agreements with the US and other countries as well as the Nuclear Non-Proliferation Treaty. Thus, Japan is restricted from nuclear armament by domestic and international laws.

The US would not readily agree to Japan's nuclear armament as it would provoke Japan's neighbours such as South Korea, North Korea, and China and might lead to nuclear proliferation in the region. South and North Korea, China, and Russia would disapprove and Japan would draw criticism from the rest of the world as well. In other words, it is not realistic to think that Japan could have a nuclear deterrent on its own. It would be an extremely dangerous move that would only pave the way to nuclear war. The nuclear deterrence theory is unrealistic especially in Japan.

Reasons to keep nuclear power plants

Why can't Japan relinquish its nuclear power plants?

The Fukushima accident has clearly shown that there is no such thing as absolute safety in terms of nuclear power generation and that risks remain very high for an earthquake-prone country like Japan. It has also demonstrated that electricity supply is not adversely affected by the absence of nuclear energy, as all nuclear power plants have been off line for twenty-three months between 2013 and 2015.

Electric power companies are afraid that if they decide to change their business plan to zero nuclear energy, they will suffer an extraordinary loss as they lose asset values in terms of their nuclear power generation plants, equipment and nuclear fuel. If the extraordinary loss exceeds a company's net asset value, the company technically becomes insolvent and fails. As an extraordinary loss cannot be included in the costs used to calculate electricity rates, the companies cannot cover the loss by way of higher electricity charges. Thus, the electric power companies that are highly dependent on nuclear power are against denuclearization for fear of corporate failure (Genshiryoku Shimin Iinkai 2014: 198–199).

In order to mitigate such effects, changes to the current accounting system will be needed, including allowing the depreciation of nuclear power generation plants for a certain period after abolishment. In fact, some changes were made to the accounting system in October 2013. We need to work out a transparent mechanism with a clear division of responsibilities between the national government and electric power companies in order to avoid confusion about how to treat insolvency and business failure.

Towards a post-nuclear future

Which direction is the way for Japan to survive as a leading technology state in the 21st century? Keeping nuclear power generation, or promoting technology for efficient use of energy and renewables? I hope that 11 March 2011, the first day of the Fukushima disaster, will be the turning point of Japanese society, East Asia, and the world, to shift energy policy towards a post-nuclear future.

References

Beck, Ulrich et al. (eds) (2011), *Risuku ka suru Nihon shakai* [Japanese risk society], Tokyo: Iwanami Shoten.

Genshiryoku Shimin Iinkai (2014), *Genpatsu zero shakai heno michi* [Our path to a nuclear-free Japan], accessed 31 January 2017 at www.ccnejapan.com/?page_id=3000.

Hasegawa, Koichi (2012), Facing nuclear risks: Lessons from the Fukushima nuclear disaster. *International Journal of Japanese Sociology*, 21, 84–91.

Hasegawa, Koichi (2014), The Fukushima nuclear accident and Japan's civil society: Context, reactions, and policy impacts. *International Sociology*, 29(4), 283–301.

Hasegawa, Koichi (2015), *Beyond Fukushima: Toward a post-nuclear society*, Melbourne: Trans Pacific Press.

Hirose, Hirotada (2013), Fukushima daiichi genpatsusaigai wo miru seron [Public Opinion on the Fukushima Nuclear Disaster]. *Kagaku (Science Journal)*, 83(12), 1346–1353.

ICANPS (Investigation Committee on the Accident at the Fukushima Nuclear Power Stations of Tokyo Electric Power Company) (2012), *Final Report*, accessed 31 January 2017 at www.cas.go.jp/jp/seisaku/icanps/eng/.

Independent Investigation Commission on the Fukushima Nuclear Accident (2014), *The Fukushima Daiichi nuclear power station sisaster: Investigating the myth and reality*, New York: Routledge.

Ishibashi, Katsuhiko (2011), Masani 'genpatsu shinsai' da [A 'combined earthquake and nuclear disaster' has finally happened], *Sekai (The world)*, 817, 126–133.

Iwai, Noriko and Kuniaki Shishido (2015), The impact of the great east Japan earthquake and Fukushima Daiichi nuclear accident on people's perception of disaster risks and attitudes toward nuclear energy policy. *Asian Journal for Public Opinion Research*, 2–3, 172–195. doi:10.15206/ajpor.2015.2.3.172.

Jomaru Yoichi (2012) *Genpatsu to media: Shimbun journalism nidome no haiboku* [Nuclear Energy and Media: The Second Defeat of Newspaper Journalism]. Tokyo: Asahi Shimbun Syuppan.

Lotan Gilad et al. (2011) The revolutions were tweeted: Information flows during the 2011 Tunisian and Egyptian revolutions. *International Journal of Communications*, 5: 1375–1405.

NAIIC (National Diet of Japan Fukushima Nuclear Accident Independent Investigation Commission) (2012), *The National Diet of Japan: The official report of the Fukushima Nuclear Accident Independent Investigation Commission*, accessed 31 January 2017 at http://f-archive.jaea.go.jp/handle/faa/115/warp.da.ndl.

Shimizu, Shūji (2011a), Genshi ryoku zaisei wo kokumin no te ni [Put nuclear finance back into the hands of citizens]. *Kagaku (Science Journal)*, 81(7), 676–679.

Shimizu, Shūji (2011b), Dengen sanpō ha haishi subekidearu [Three laws for power facility development should be abolished]. *Sekai (The world)*, 819, 96–103.

Soeda, Takashi (2014), *Genpatsu to ōtsunami keikoku wo hōmutta hitobito* [Nuclear power stations and tsunami disasters: they are responsible for the incidents], Tokyo: Iwanami Shoten.

2 The anti-nuclear movement and ecological democracy in South Korea

Dowan Ku

I Introduction

What can we learn from the Fukushima nuclear disaster? We realized that we share a common nuclear risk all over the world. We cannot help reflecting upon modern technology's omnipresent risk. This risk is not confined to nation states and the present generation. Global risk at present is not bounded by time and space.

How can we imagine the post-Fukushima world? Will it be a nuclear risk society or a post nuclear society? A nuclear risk society can be defined as a society where nuclear power generation and weapons proliferate or are sustained, and nuclear risk increases or cannot be managed. A post nuclear society can be defined as a society where nuclear power generation and weapons are abandoned and the nuclear risk can be managed ecologically and democratically.[1]

On the one hand, we can see that European countries such as Germany, Switzerland, and Italy are moving to a post nuclear society. On the other hand, Asian countries such as South Korea and China continue to increase their nuclear plants. Why the difference? Economic structures, development models, energy supply and demand, and other factors could explain this difference. However, one of the most important factors in the choice between the nuclear risk or post nuclear society paths is the environmental movement, specifically the anti-nuclear movement.[2] For example, the German anti-nuclear movement and its Green Party are strong enough to change nuclear policy, while the anti-nuclear movement in Asian countries such as Korea and China, is not as strong as Germany's. We can suppose that the stronger the anti-nuclear movement is, the less nuclear power generation or armament will be developed.

This chapter tackles the following research questions. Firstly, how influential was the anti-nuclear movement in South Korea? Secondly, how can we evaluate the nuclear politics of South Korea in terms of ecological democracy? Thirdly, what is the relationship between nuclear power and democracy?

II Nuclear risk and ecological democracy

1 South Korea's nuclear risk

The nuclear risk in East Asia is high, and has increased, specifically in South Korea. South Korea hosts 25 commercial nuclear reactors, with five more under construction and another six having confirmed construction plans. Nuclear power comprised 24.8% of Korean generation facilities and 31.4% of gross generation in 2010. The Korean government plans to increase nuclear power's share of gross generation up to 48.5% by 2024 (Yun 2011). Furthermore, older nuclear power plants that have operated since 1978 (Gori reactor 1) and 1983 (Wolsung reactor 1) pose serious risks. Gori reactor 1 has experienced many accidents and is being operated beyond its design life. Korean nuclear power facilities' capacity per unit area is the second highest in the world (Yun, 2011).[3] The nuclear risk is real and high in South Korea.

2 Complex and structural risk

Nuclear power is a symbol of industrial modernity. It has two faces: nuclear power plants and nuclear weapons. Nuclear power poses serious risks from radioactive

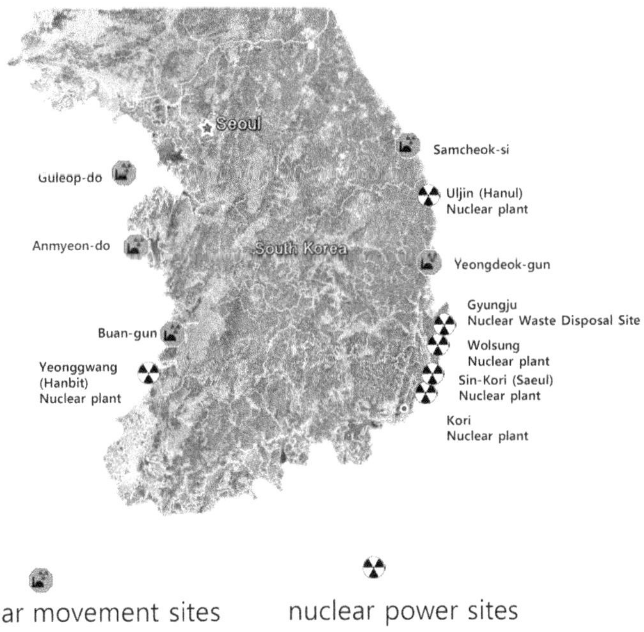

Figure 2.1 Sites of nuclear facilities and anti-nuclear movement

contamination. How has it proliferated worldwide despite this risk? We suggest structural factors such as nationalism and industrialism.

Firstly, nationalism has a strong relationship with nuclear power. Nuclear power plants and weapons were made for military purposes by both imperialist states and others. Nuclear plants can produce the raw materials for nuclear bombs. That is one reason why Japan, South and North Korea, and other countries have continued building and operating nuclear power plants.[4]

Secondly, nuclear power has been the base of industrialism, because it can support mass production and consumption of electricity. Most capitalist and socialist industrial countries are dependent on fossil fuels and nuclear power for economic growth. Countries which are suffering from shortages of fossil fuel resources are eager to sustain or expand nuclear power plants.[5] We can say that nuclear risk is a structural risk based on nationalist antagonism and industrialism stemming from technological optimism.

In East Asia, along with these two factors, we can find another factor, which is the 'developmental state' or 'developmentalism'.[6] Developmentalism can be defined as the strategy and discourse in which most resources are mobilized for national economic growth. In this discourse, both authoritarian and democratic regimes can be legitimized by economic growth driven by the state.

Nuclear power has proliferated in developmental states such as Japan, Taiwan, South Korea, and China. Authoritarian or conservative governments could mobilize various resources to build up nuclear power plants and suppress protests by local residents and environmental groups. In this regard, developmentalism has a strong relationship with nuclear power. Elites argued that nuclear power was essential for economic growth and national prosperity. Sacrificing the few for the many was legitimized in the discourse of national prosperity and industrial development. Antinuclear movements were suppressed under authoritarian regimes in South Korea, Taiwan, and other places. We can summarize that the East Asian nuclear risk is a complex result of nationalism, industrialism, and developmentalism.

3 Decision makers of nuclear risk: nuclear power alliance

Who decided South Korea should take nuclear risks? The first initiator of nuclear power projects in South Korea was the United States government. The US government started to develop nuclear power plants and tried to create an international coalition around 'atoms for peace' in order to build an anti-communist alliance. A South Korea–US nuclear cooperation agreement, initiated by the US government, was signed in 1955 and ratified in 1957. An act on atomic power was legislated in 1958 and the Korean Atomic Institute was established in 1959. Nuclear power plants could not be constructed in the 1960s for technological and economic reasons, particularly because they were not as economical as hydroelectric power (Yun and Oh 2006).

However, nuclear policy changed dramatically in the 1970s. The first factor was a nuclear weapon development project and the second was the oil crises of 1973 and 1979. The US government decided to reduce American troops in South Korea in 1970. President Park Jung-hee secretly started a nuclear weapon development project in 1971, though it was not successful because of the anti-nuclear proliferation policy of the US government. On the other hand, the oil crisis in 1973 provided a good opportunity for nuclear power plant construction. The Korean government decided to develop nuclear power plant technology for 'peaceful use'. The Korean government and technocrats had a strategy for independent nuclear technology development from an early date.

The Korean nuclear power plants were developed fast and sustainably in the late 1970s and 1980s, under the electric power public enterprise system. KEPCO (Korean Electric Power Company) merged a nuclear power facility firm and research institute into affiliated companies in the early 1980s and a strategy of nuclear power standardization and localization for Korean type nuclear power was implemented on the basis of a public company system.[7] This resulted in a system of low-priced electric power and encouraged electrification. The myth that nuclear power is cheap was reinforced by this system (Hong 2016). The nuclear power industry succeeded in exporting a plant to the United Arab Emirates (UAE) in 2009, with strong support from the government.

In sum, a nuclear power policy was initiated by President Park, who was eager to develop nuclear weapons. This project was frustrated by the US government. On the other hand, the oil crisis in the 1970s awoke the energy security issue and the necessity of energy options other than oil. In this context, nuclear power plants were highlighted as economical, clean, and high-tech facilities. The risks of nuclear power were concealed and transferred to the people living near nuclear facilities, as well as to future generations and non-human beings. Bureaucrats in the economic, industrial, energy, science, and technology sectors built a concrete nuclear power alliance, although they competed for their own initiatives and interests. The alliance was consolidated by the public electric power production and distribution system. Construction companies could join the nuclear power alliance as strong partners. This alliance could mobilize strong support from politicians and mass media. 'Nuclear Mafia' describes the hidden connections among nuclear power interests. Major participants in this nuclear alliance included nuclear power companies, economic bureaucrats, the energy and science sector, construction companies, pro-nuclear politicians, and the conservative mass media.

The characteristics of the Korean nuclear power alliance can be summarized as follows. Firstly, the alliance was initiated by the failed nuclear weapon development strategy. Secondly, it was established for the sake of energy security after the oil crises in the 1970s. Thirdly, it was consolidated by the public electric power system, which coordinated groups such as construction, industry, scientists, and technicians. Fourth, once this alliance was constructed, it succeeded in standardizing and localizing nuclear power. Fifth, this alliance was also successful in mobilizing a nuclear power advocacy coalition that included politicians, conservative mass media, and local residents near nuclear facilities.

The nuclear alliance was tackled by a strong anti-nuclear movement from the late 1980s through to the early 2000s. Though the alliance could co-opt many local residents and exclude anti-nuclear activists, they were challenged by a new anti-nuclear movement after the Fukushima accident.

4 Structure of nuclear risk: developmentalism and authoritarianism

Developmentalism is a combination of nationalism, industrialism, and capitalism. Political leaders and bureaucrats in South Korea mobilized people for national industrialization and excluded labour unions and social movement organizations from public participation before 1987. This kind of mobilization system was justified as promoting national security and independent development. Nuclear power plants have proliferated under the context of developmentalism in South Korea.

A nuclear risk society under developmentalism is ecologically unsustainable and socially unjust. How can we move from this nuclear risk society? In order to find the answer, we need to talk about 'democracy', specifically 'ecological democracy'. Table 2.1 categorizes four types of governance concerning developmentalism and democracy.

Firstly, we can divide political regimes between authoritarianism and democracy. The former is a regime which is closed and the latter is open to the public. An open public sphere, equitable quality of life, and participatory decision-making processes are among the main characteristics of democratic regimes.

Traditionally, political regimes were the domain of social scientists. However, when we witness global environmental problems such as climate change and nuclear risk, we should add a new dimension, which is ecologism. Therefore, we can divide regimes among developmentalism and ecologism in terms of development and ecological conservation. Developmentalism is related to industrialism, capitalism, and technological optimism. Ecologism is a discourse in which ecological balance, conservation, and sustainability are prioritized over economic growth and development. Ecologism is critical to anthropocentrism and human exemptionalism (Catton and Dunlap, 1980). When we add ecologism to the types of governance, we find two types of ecologism: ecological authoritarianism and ecological democracy.

These four types of governance have different positions on nuclear power. Developmentalism needs sufficient and efficient electricity. Nuclear power is an essential resource for developers. Both developmental authoritarians and developmental democrats agree on nuclear power, the 'peaceful use of atomic power'. Ecological authoritarians are ambivalent on the nuclear issue. Some 'ecologists' such as James Lovelock urged nuclear power as the solution for climate change (2006). Ecologists, who believe in technological fixes, may support nuclear power.[8] However, ecological democrats have a tendency to oppose nuclear power. Why? Here we need to think about the theory of ecological democracy.

Table 2.1 Four types of governance concerning development and authoritarianism

	Authoritarianism	*Democracy*
Developmentalism	Developmental Authoritarianism	Developmental Democracy
Ecologism	Ecological Authoritarianism	Ecological Democracy

5 Discourse for a post nuclear society: ecological democracy

Nuclear power plants have increased under the governance of developmental authoritarianism in South Korea. Nuclear power plant construction was initiated by the Park Jung-hee regime. President Park was eager to construct nuclear power plants for electricity and weapons. There was little protest against nuclear power plants under developmental authoritarianism from 1961 to 1987. However, the anti-nuclear movement flourished since democratization in 1987. In this regard, we need to think about the relationship between nuclear risk and democracy. Can we move to a post nuclear society if we can develop democracy?

Democracy as rule by a majority of the people has been the modern political norm. However it is always threatened by the capitalist and bureaucratic state system. In global neo-liberalism, democracy manifests the limitations of addressing social polarization and exclusion. In this context, deliberative democracy has been discussed as an alternative to liberal democracy. According to Habermas (1989), communicative rationality can overcome dogmatic rationality in ideal speech situations. Following Habermas, John S. Dryzek researched ecological democracy or green democracy beyond deliberative democracy, and suggested discursive democracy, which 'can embrace difference as well as consensus, the public sphere as well as the state, transnational as well as domestic politics, and nature as well as humanity' (Dryzek 2000: 175).

This 'deliberative turn' can be a promise of ecological sustainability, because the more deliberation and participation, the greater the possibility of deliberation about future generations and non-human beings. However, this theoretical discussion is based on normative political theory. As Chantal Mouffe said, real politics are motivated by 'power and antagonism' (Mouffe 1999: 752). In real politics, power elites limit their sovereignty only when their hegemony (a combination of consent and force) is in danger. A power sharing state can be found only when political and social resources can be shared by adversaries. Therefore, we need to analyse the procedure and substance of real environmental politics.

Green democracy or ecological democracy in South Korea has been discussed in its theoretical and empirical dimensions. Soonhong Mun researched the relationship of democracy and ecology and suggested creating a green public sphere and making the public sphere green. Sun-Jin Yun (2007) analysed the limits of liberal democracy in the case of a local referendum regarding nuclear waste dump sites and argued that ecological justice should be achieved. These researchers were all interested in the advantages as well as disadvantages of deliberative democracy, however they could not find successful cases combining deliberative democracy procedures with ecological substance. Parallel progress of democracy and ecology may be possible and desirable, but it is difficult in the real world, where industrialist capital and state bureaucrats have the dominant power.

Ecological democracy can be defined as 'the political culture and practice in which the rights of social minorities, future generations, and non-human beings can be deliberated and realized in the open and participatory public sphere and policy process'. Ecological democracy cannot be achieved by deliberation alone.

Public deliberation on ecology can be prosecuted only with a powerful green public sphere. How can a green public sphere be constructed? It can be constituted by active discussion and deliberation initiated by environmental movements, specifically by direct action. In this sense, environmental movements can be a key factor that can determine whether we have a nuclear risk society or a post nuclear society.

6 Evaluation of nuclear politics in terms of ecological democracy

How can we evaluate social phenomena in terms of ecological democracy? Ecological democrats focus on the distribution of ecological risk in time and agency. We can raise two key issues of ecological democracy:

1 Equitable distribution of ecological risk among socio-ecological minorities, future generations, and non-human beings.
2 Participation and deliberation of socio-ecological minorities, representatives of future generations, and non-human beings.

The first is a substantial ecological justice issue and the second is a procedural issue. In terms of these two issues, how can we evaluate the Korean anti-nuclear movement and nuclear policies? Six analytical questions can be raised in order to answer this question.

Are the rights of socio-ecological minorities protected?
Are the rights of future generations protected?
Are the rights of non-human beings protected?
Is there a participatory policy process for socio-ecological minorities?
Is there a participatory policy process for future generations?
Is there a participatory policy process for non-human beings?

If we can find six positive answers for these questions when we analyse cases, regimes, or politics, then we can say that they are ecologically democratic cases. For example, the Kim Daejung administration decided to cancel the Donggang Dam construction project in favour of ecological conservation and future generations under a participatory policy process in 2000. We can evaluate that case as an ecologically democratic policy process and implementation. The remainder of this chapter will use these analytical questions to analyse the anti-nuclear movement and nuclear politics in South Korea. Specifically, it will analyse the relationship between the anti-nuclear movement and pro-nuclear policies.

III The anti-nuclear movement before Fukushima

South Korea had a strong anti-nuclear movement among local residents and environmental NGOs until 2004. The government could not find new sites for nuclear waste disposal and power plants. However, the movements were not powerful enough to change national nuclear policy. After the government changed its policy for nuclear waste disposal in 2005, the anti-nuclear movement quickly weakened. As the nuclear movement retreated, the Lee Myung-bak administration implemented nuclear plant enlargement and export policies. This section provides an overview of the history of the anti-nuclear movement.[9]

1 The beginning of the anti-nuclear movement (ANM: 1987–1989)

The Korean anti-nuclear movement began with a local residents' movement around the Yeonggwang nuclear plant in Jeollanam-do province. Fishermen around the Yeonggwang nuclear plant organized protests in 1987 against fisheries damaged by hot drain water from the plant. In addition, illegally dumped nuclear waste was found in 1988 around Kori in Gyeongsangnam-do province. Local residents near Kori, Yeonggwang, and the Wolsung plant in Gyeongsangbuk-do province organized anti-nuclear demonstrations in 1988. On the basis of these efforts, the National Headquarters for the Nuclear Plant Banishment Movement was organized in 1989 by several environmental health movement organizations and local residents committees.

During this period, the rights of socio-economic minorities, future generations, and non-human beings were not deliberated, and no policy process for these was performed at all. Developmental authoritarianism was dominant during this period.

2 Success of the anti-nuclear waste disposal movement (1990–2004)

Compared with the anti-nuclear plant movement, the anti-nuclear waste disposal facility campaign was more intense and influential. Anmyondo (Chungcheongnam-do province, 1990), Gulupdo (Incheon-si, 1994), and Buan (Jeollabuk-do province, 2003–2004) were designated as middle and low level radioactive waste disposal sites by the government. The residents of these areas strongly resisted against the government, carrying out direct actions such as sit-ins, demonstrations, and school boycotts.

The Buan case was unique. Residents in Buan organized cultural performances and strong resistance against the police. They successfully carried out a voluntary referendum among residents on the basis of their unified power. After the success of their campaign, some leaders began developing an energy independent village to exemplify a nuclear free society.

Although local movements were successful in stopping nuclear waste disposal, nuclear reactors were continuously constructed. Because the government and KEPCO could not find new plant sites after the 1990s, they constructed and are constructing new reactors on the existing plant sites. Five reactors are under construction near the existing plants. Residents around nuclear plants could not resist against the additional risk of constructing new reactors. It was hard for them to overcome institutionalized compensation and the economic benefits of nuclear plants. Most residents and municipalities have been subjected to the nuclear alliance of the central government, KEPCO, big businesses, and conservative mass media.

The anti-nuclear movement was successful in stopping nuclear waste disposal and new designation of nuclear plant sites at the local level until 2004. As local anti-nuclear waste disposal facilities campaigns were developed in the 1990s, national solidarity for changing nuclear power policy was also activated. However, the national anti-nuclear movement against nuclear power policy was not so well developed. Anti-nuclear efforts were confined to local issues at the specific sites which were designated as nuclear waste disposal facilities.

Although anti-nuclear activists used the anti-nuclear waste disposal campaign as political leverage for phasing out nuclear plants, they could not change national nuclear policy. No national mass anti-nuclear movement developed and the policy window was not opened. While national environmental NGOs suggested a round table for deliberating a reevaluation of nuclear policy to the Roh Moo-hyun administration in 2004, this was not conducted.[10]

On the other hand, nationwide environmental NGOs started a new movement for energy alternatives in the 2000s. They advocated constructing energy independent villages, alternative energy policies, and citizens' solar power plants. For example, the Korean Federation for Environmental Movement (KFEM) launched a new sub organization, 'Center for Alternative Energy (CAE)', for the alternative energy movement in 2001. CAE became an independent NGO in the mid-2000s. This organization initiated citizens' solar power plants and suggested many options for renewable energy and a post nuclear society.

A community-based alternative energy movement has also developed since the mid-2000s. After the anti-nuclear waste disposal facility campaign in 2004 at Buan, some leaders of the campaign such as Catholic Father Mun Gyu-hyun and citizen Lee Hyun-min initiated an energy independent village project in a small village in Buan-gun called Deungyong maul. The anti-nuclear campaign at Buan was motivated by anti-risk and 'keep our community' sentiments. The campaign succeeded by implementing and winning a voluntary residents' referendum in 2004. Some leaders tried to go beyond 'Not In My Back Yard (NIMBY)' sentiment to initiate ecologically sustainable autonomy from the bottom. They established solar power plants and biomass facilities and initiated energy conservation projects in the villages.

During this period, the rights of socio-economic minorities were deliberated and improved, step by step. Government passively reformed and increased subsidies for the residents around nuclear power plants and waste disposal sites. However, it is hard to say that the rights of minorities really improved. The increase in subsidies cannot reduce the risk facing the residents around nuclear power plants and disposal sites. There is no evidence of improvement in the rights of future generations and non-human beings. Nuclear risk increased in this period because of new reactors' operation. The anti-nuclear movement could not interrupt new reactor construction and operation.

The participation of local residents and environmental NGOs and intellectuals in the policy process during this period was more developed than before. However, it was still very limited and passive. No participatory and deliberative policy process for future generations and non-human beings were developed during this period, in spite of democratization. Political democratization after 1987 could not extend to the public sphere for a post nuclear society.

3 Retreat of the anti-nuclear movement (2005–2007)

The Buan anti-nuclear campaign was successful in blocking nuclear waste disposal, but not as successful in changing nuclear waste disposal and generation policy. After the government failed to determine middle and low level radioactive waste disposal facilities, it initiated a new policy in 2004. They separated low and medium level radioactive waste from high level waste and confirmed they would not construct high level radioactive facilities in the low and medium facility sites. Furthermore, they promised enormous economic subsidies.[11] Because of this aggressive policy change, many local governments and residents wanted to host the facility. The government decided to award the facility to the local authorities that won the highest agreement in a residents' referendum.

This policy change totally transformed the political arrangement of nuclear issues. The nuclear waste disposal facility changed from a NIMBY (Not In My Back Yard) facility to a PIMFY (Please In My Front Yard) facility. Ultimately, Gyungju in Gyungsangbuk-do province, where the anti-nuclear movement was weakest, was designated as the facility site. In this situation, the nuclear movement rapidly weakened. It could not respond to this very different social context. The former risk frame was dramatically changed to a 'money frame' by a well organized government strategy.

During this period, the rights of minorities were partially improved. Subsidies for nuclear waste dump sites were greatly increased and local governments' and residents' participation in the policy process was partially improved under the Roh administration. The trade off of money and risk occurred under a partial, even pseudo, participation process. However, the nuclear risk to future generations and non-human beings was not decreased. Furthermore, deliberative and participatory policy processes for an ecologically sustainable future were seldom performed.

4 Nuclear power renaissance (2008–2010)

On the basis of the pro-nuclear policy success of the Roh administration, the Lee Myung-bak administration carried out a nuclear power enlargement and export policy. The Lee administration declared 'Low Carbon Green Growth Policy' as a national development policy in 2008. In this policy programme, nuclear plant enlargement and export were central. They argued that nuclear plant enlargement was safe and good for tackling climate change. The Lee administration contracted with the UAE to export a Korean-style nuclear plant in 2009. This contract provoked a nuclear plant export competition between Korea and Japan.

With this nuclear enlargement and export policy, the nuclear risk to minorities and future generations and non-human beings around the world is increasing. No deliberative and participatory policy process for risk reduction and management was invoked in this period. Developmental authoritarianism was revitalized after 2008.

5 Anti-nuclear movement after Fukushima (2011– present)

The Fukushima disaster triggered new waves of anti-nuclear sentiment in Korea. Most of the media reported the accident, and it provoked serious mass concern about nuclear risks. Firstly, new nuclear reactor construction was an important issue in a local governor special election on 27 April 2011. After the Fukushima disaster, the ruling Hannarah party's gubernatorial candidate in Gangwon province, Um Gi-young, changed his position on nuclear power plant construction in that province from agreement to reservation. On the other hand, the candidate of the opposition Democratic Party Choi Mun-sun, opposed the invitation, and he won the election for gubernatorial.

Secondly, much discussion about a post-nuclear power path has occurred in the public sphere. Nuclear power plant safety, public health, East Asian nuclear risk, and the potential victims of the Fukushima disaster were among the main issues. Specifically, a strategy for a nuclear free Korea also has been discussed in the public sphere, including by politicians. For example, the Energy Alternative Forum, which includes opposition party members of the National Assembly, alternative energy professionals, and NGO leaders, was created in 2011 to promote nuclear free alternatives.

Thirdly, many demonstrations against nuclear power were organized by environmental NGOs, citizens, and religious leaders. Environmental NGOs focused on shutting down the Kori No. 1 reactor and Wolsung No. 1 reactor, which were old and risky. Many local citizens and environmental groups made every effort to close the two reactors. Finally, the Park Geun-hye administration decided in 2015 not to extend the operation period of Kori No. 1 reactor. Furthermore the local court in Seoul ruled in 2017 that the re-operation of Wolsung No.1 reactor was illegal and should be cancelled.

Nuclear enlargement policy was strongly influenced by the Fukushima disaster. However, the number of media reports about Fukushima decreased after April 2011. Furthermore, officials argued that Korean nuclear plants were safe and nuclear generation was optimal for meeting electricity demand and coping with climate change. The Korean Hydro & Nuclear Power Company (KHNPCO) proposed Samcheok and Yeongduck as new nuclear power plant sites in December 2011. National NGOs campaigned against this decision. Gangwon-do province governor Choi Mun-sun expressed a critical position on this decision by KEPCO.

Anti-nuclear activists in Samcheok organized a voluntary residents' referendum in 2014 to ask the opinions of residents about construction of a nuclear power plant there, without permission from the central government. In the referendum, 68% of eligible voters participated and 85% of the voters were opposed to the nuclear plant construction. Activists in Yeongdeok county, which is the other proposed site for a new nuclear power plant, also organized a residents' referendum in 2015. In this referendum, 33% of eligible voters participated and 92% of the voters were opposed to the nuclear plant construction. Although the central government denied the legal validity of both votes, it may be very difficult for the government to construct nuclear power plants in Samcheok and Yeongdeok.

The Fukushima accident revitalized the anti-nuclear movement and mobilized new concerns and new citizens who are sensitive to nuclear risk, such as housewives in consumers' co-operatives. 'The group of professors for post-nuclear power' was organized in November 2011. A group of lawyers for post-nuclear power, 'Haebaragi (Sunflower)', was also organized in February 2012.

The main opposition party, the Democratic Party, declared in December 2011 that they would reevaluate the nuclear plants in terms of sustainability and human peace. The newly elected mayor of Seoul Metropolitan City, Park Won-sun, declared in January 2012 that he would decrease 2 million TOE (Tonnage of oil equivalent) by 2014 by implementing a strong energy conservation and renewable energy policy. The 'reduction of one nuclear power reactor policy' through energy conservation was very successful. In addition, 'the group of local government heads for post-nuclear power and energy conversion' was organized in February 2012. The group declared that it would try to conserve energy, shut down old reactors, and oppose construction of new nuclear power plants.

Thousands of citizens had an open-air meeting for post-nuclear power in front of Seoul Metropolitan city hall in memory of the Fukushima disaster in March 2012. It was hard to imagine these kinds of paradigmatic changes in the anti-nuclear movement before the Fukushima accident. These changes are a meaningful sign of new political terrain. We can find a cleavage in the pro-nuclear coalition after Fukushima.

On the other hand, many environmentalists who suffered from the retreat of the environmental and anti-nuclear movements have been trying to build a Korean Green Party. The Korean Green Party was founded in March 2012.

In this period, the anti-nuclear movement has changed its focus from local issues to national and global issues, and from short-term issues to long-term issues. Members of civil society, specifically housewives and intellectuals such as professors and lawyers, created their own organizations to promote a nuclear free future. The most important change is that local government heads have announced post-nuclear power policies. We can find a new ecological public sphere for a post-nuclear society in civil society and local governments.

During this period, the nuclear risk facing minorities, future generations, and non-human beings did not decrease. New policy processes were not performed. However, we can find the very beginning of a structural change among civil society and local governments towards a new risk reduction and policy process.

IV Discussion

Let us discuss the three research questions on the basis of the history of the anti-nuclear movement and nuclear politics. Firstly, how influential was the anti-nuclear movement in South Korea? We could find that the anti-nuclear movement developed since 1987 and flourished from 1990 to 2004. During this period, new nuclear power plant and waste disposal sites could not be designated. Local residents and environmental NGOs in a strong anti-nuclear movement could defer the designation of new nuclear facilities. The government could only construct new reactors around old power plant sites, where the anti-nuclear movement was not as strong. The anti-nuclear movement could exercise its influence to defer the new location of plants, but it could not change nuclear power policy at the national level.

However, the new nuclear waste disposal policy of 2005 changed the political terrain of the nuclear issue. Many local government heads tried to invite nuclear disposal facilities and power plants. The Lee Myung-bak administration could implement nuclear power plant enlargement and export policy in a context of weak anti-nuclear movements. We can conclude that nuclear power plant enlargement can be sustained when the anti-nuclear movement among local residents and environmental groups is weak.

Secondly, how can we evaluate the history of the anti-nuclear movement and politics in terms of ecological democracy? Nuclear politics in South Korea can be evaluated in terms of ecological democracy as shown in Table 2.2.

We can summarize that the right and participation of socio-ecological minorities improved from 1990 to 2004 through the direct action of grassroots movements, and that ecological democracy was partially developed in that period. However, the rights and participation of future generations and non-human beings were not improved even then. Ecological democracy was developed only at the local level and the present generation level.

Table 2.2 Evaluation of Nuclear Politics in terms of Ecological Democracy

Years	*1987–89*	*1990–2004*	*2005–2007*	*2008–2010*	*2011–Present*
Periods	Beginning of ANM	Success of ANM	Retreat of ANM	Nuclear Power Renaissance	New Beginning of ANM
Rights of Minorities	Not protected	Self-protected by grassroots movements	Trade off of money and risk	Risk increased	Protected by residents, citizens, and politicians
Rights of Future Generations	Not protected	Protected at the local level	Not protected	Risk increased	Risk slightly reduced
Rights of Non Human beings	Not protected	Protected at the local level	Not protected	Risk increased	Risk slightly reduced
Participation of Minorities	Excluded	Informal participation by direct action	Partial and pseudo participation	Excluded	Partially included by local government
Participation of Future Generations	Excluded	Excluded	Excluded	Excluded	Excluded
Participation of Non-human Beings	Excluded	Excluded	Excluded	Excluded	Excluded

This kind of weak ecological democracy retreated after 2005. The government and the pro-nuclear alliance co-opted politically and economically weak local governments and residents. They exposed local residents, future generations, and non-human beings to high nuclear risk. Weak ecological democracy for present generations was replaced by developmental authoritarianism after 2008.

Thirdly, what is the relationship between nuclear power and democracy? If political democratization is successful, can a nuclear free society or post nuclear society be possible? In the Korean case, the answer is negative. South Korea has been democratized since 1987. The Kim Dae-jung (1998–2002) and Roh Moo-hyun (2003–2007) administrations developed political democratization in terms of institutions. However, they did not create a public sphere for discussing and deliberating alternative energy policy and phasing out nuclear power. The Roh administration suppressed the anti-nuclear waste dump campaign in Buan and did not initiate public discussion on nuclear power and waste disposal policies. They initiated a new policy to increase subsidies in order to designate middle and low level radioactive waste disposal sites. Money was offered to offset risks in place of precautionary risk management. The Roh administration tried to demonstrate the government's power to solve social conflict over nuclear issues on the one hand, and still supply sufficient electricity for economic development. Economic development and political legitimization were the imperatives, both under authoritarian and democratic regimes.

We can conclude that although political democratization has occurred in South Korea, as long as the government's imperative of developmentalism is sustained, a post nuclear society cannot be attained. We can call the Kim Dae-jung and Roh Moo-hyun administrations governments of 'developmental democracy'. Under developmental democratic governance, the nuclear risk of socio-economic minorities can be partially paid off by subsidies. However, the nuclear risk was then shifted to future generations and non-human beings. Furthermore, deliberation and participation was not carried out well on the nuclear issue. In conclusion, ecological democracy was not developed during the 'democratic administrations' of South Korea.

The Lee Myung-bak administration weakened democratic institutions in many areas. For example, the environmental impact assessment and prior feasibility review for huge construction projects were neglected or not performed well, as in the case of 'Four major river project', which destroyed a natural river ecosystem. The Lee administration implemented a nuclear power enlargement and export policy called 'nuclear power renaissance'. Under the Lee administration, it is hard to find a green public sphere where discussion and deliberation on an ecologically sustainable post nuclear society can occur. The nuclear risk to minorities, future generations, and non-human beings was increased and the deliberative democratic procedure was destroyed. We can call the Lee administration 'developmental authoritarianism'.

South Korea experienced a path of developmental authoritarianism, then developmental democracy, then a return to developmental authoritarianism. The environmental and anti-nuclear movements flourished in the 1990s, but their influence in the public sphere and institutions decreased since the mid-2000s. Neo-liberal developmentalism dominated the precautionary green public sphere.

This socio-political context contrasts with the German case. Environmental and anti-nuclear movements and green politics were well developed under the passively exclusive German state (Dryzek *et al.* 2003). The strong anti-nuclear movement, Green Party, and well-organized research institutes for alternative energy could change nuclear power policy. There we can find some characteristics of 'ecological democracy' in which risk and the well-being of social minorities, future generations, and non human beings could be deliberated and the representatives of these agencies could participate. The Germans are on an 'ecological democracy path based on a strong environmental movement'. This contrasts with Korea's 'developmental authoritarianism path based on a weak environmental movement'.

V Conclusion

This chapter discussed the anti-nuclear movement and ecological democracy in South Korea. The essential forces of ecological democracy are environmental movements in civil society. When the anti-nuclear movement was powerful, nuclear policy could not be implemented well. However, the movement was not strong enough to fundamentally change national nuclear policy. Although the anti-nuclear movements were successful in local areas, they were not successful at changing nuclear policy or building a post nuclear society. We can conclude that a post nuclear society can only be built on a strong anti-nuclear movement.

Secondly, it evaluated environmental politics in terms of ecological democracy. When the anti-nuclear movement was powerful, weak and short-term ecological democracy for socio-ecological minorities improved, but when the anti-nuclear movement was dominated by the government after 2005, ecological democracy was replaced by developmental authoritarianism.

Thirdly, nuclear power has a strong relationship with developmentalism and authoritarianism. It advanced under developmental authoritarianism in the 1970s and 1980s. Democratization opened policy windows for a post nuclear society. However, democracy itself was not enough for phasing out nuclear power generation because the democratic regime was dominated by developmentalism. Under democratic democracy, the public sphere for a post nuclear society was not well constructed. Korea experienced a 'U-turn path to developmental authoritarianism' instead of moving to the 'ecological democracy path'. Democracy should be ecologized to produce a post nuclear society.

However, the Fukushima disaster provoked reflection on the fragile characteristics of industrial modernity and developmentalism. We can find new green public spheres for a post nuclear and nuclear free society in Korea, Japan, Taiwan, and the rest of the world. This green public sphere based on environmental movements will open new horizons for ecological democracy.

Notes

1 A nuclear free society is a society where the ecosystem is free from anthropogenic nuclear risk. It is hard to anticipate a nuclear free society, because radiation from spent nuclear fuel and high level nuclear waste lasts more than thousands of years.
2 In this article, the research object is the anti-nuclear power generation movement, not the anti-nuclear weapon movement.
3 The highest is Belgium (195kW/km2).
4 Nuclear weapons and nation state antagonism are not the main subjects of this article.
5 Technological optimists overestimate the technological solution and underestimate the risk.
6 The developmental state is a concept developed by Chalmers Johnson (1982). He emphasized the bureaucratic autonomy for national economic growth and efficient industrial policies of the government of Japan. South Korea and Taiwan are the other models of the developmental state.
7 Korean construction companies could participate in nuclear power and radioactive waste facilities' construction facilities.
8 Here, we can find ecological authoritarianism and survivalism. Neo-Malthusians emphasize the imminent ecological crisis and inevitability of an authoritarian solution by elites and technocrats (Dryzek 2013)
9 The anti-nuclear power generation movement in Korea has two major inter-related issues: the construction of nuclear plants and nuclear waste disposal facilities.
10 The Presidential Commission on Sustainable Development of the Roh Moo-hyun administration researched an alternative sustainable energy policy, but it could not transform the nuclear energy policy frame.
11 Government promised to supply more than 300 billion won (about 270 million dollars) and move the headquarters of Korea Hydro & Nuclear Power Company to the local area which will be designated as the disposal site.

References

Korean

Hong, Deokhwa. (2016). *The Formation and Transformation of Nuclear Industry in Korea*. Ph.D. Dissertation, Seoul National University.

Kim, Hye-jung. (2011). Role of civil society and anti-nuclear movement in Korea after Fukushima. *Citizen and World (Shimin gwa Segae)*17.

Mun, Soonhong. (2000). Reconstitution of discussion on articulation of democracy and environment: Designing ecological democratization, *Korean Political Association Journal (Hankukjeongchihakhoebo)* 34(2).

Yun, Sun-Jin. (2007). The prospect and challenges to ecological democracy: Based on the evaluation on the siting process of low- and medium-level radioactive waste disposal facilities. *ECO* 11(2), 207–245.

Yun, Sun-Jin. (2011). Nuclear generation risk society and the response of civil society, paper presented at spring conference of the Korean Association of NGO studies.

Yun, Sun-Jin and Eun-Jeong Oh. (2006). A study on the social construction of the nuclear power generation policy in Korea: Focused on the introduction processes of nuclear technology. *Journal of Environmental Policy and Administration*, 14(1), 37–74.

English

Catton, William R. and Riley E. Dunlap. (1980). A new ecological paradigm for post-exuberant sociology. *American Behavioral Scientist*, 24(1), 15–47.

Dryzek, John S. (2000). *Deliberative democracy and beyond.* New York: Oxford University Press.

Dryzek, John S. (2013). *The politics of the earth: Environmental discourses.* Oxford: Oxford University Press.

Dryzek, John S., David Downs, Christian Hunald and David Schlosberg. (2003). *Green states and social movement: Environmentalism in the United States. United Kingdom, Germany, and Norway*, Oxford: Oxford University Press.

Habermas, Jürgen. (1989). *The structural transformation of the public sphere.* (Thomas Burger, Trans.). Cambridge, MA: The MIT Press.

Johnson, Chalmers. (1982). *MITI and the Japanese miracle.* Stanford, CA: Stanford University Press.

Lovelock, James. (2006). *Revenge of Gaia.* Santa Barbara, CA: Allen Lane.

Mouffe, Chantal. (1999). Deliberative democracy or agonistic pluralism. *Social Research*, 66(3), 745–758.

3 Tri-helix energy transition in Taiwan

Kuei-Tien Chou

Foreword

In recent years, global warming and climate change have seen humanity increasingly experience a dramatic collapse in the environment, economy, society, and survival, forcing the nations of the world to begin to transition from the current economic society with its highly catastrophic use of resources, energy intense, high carbon and pollution emissions (WBGU 2011). Like the rest of the world, Taiwan is simultaneously facing extensive, large-scale, multi-faceted climate change threats along with significant challenges to the economy in the transition domestically to an environmentally sustainable, social democracy. That is to say, under vertical pressure from international climate change risk and green conventions, along with horizontal pressure from society demanding the sublation of high energy consumption and industries with high levels of pollution and transition to an economy which would be more environmentally sustainable. And such a move is occurring in places all around the world; making it not just an issue of cosmopolitan governance (Beck 2006, 2010, 2015), but more of cosmopolitan risk governance (Chou 2015a), and this is happening in places around the globe including Taiwan.

Taiwan's energy transition in different in many ways to other countries; in comparison to other countries such as Germany who began this process of energy transition in the early 1980s in response to civil society strong, anti-nuclear, protest movement and protests against the decentralization of renewable energy sources, Taiwan's current energy transition simultaneously faces many other challenges including debate over energy fuel options (including reducing CO_2 emissions, decreasing use of fossil fuels, anti-nuclear dispute, and the push towards renewable energy sources), industry transition (including the sublation of industries both high in energy consumption and pollution and a move towards a low carbon economy), anti-PM2.5 Air Pollution movement, and this complex web of issues has been described by this writer as the triple helix movement of energy, industry, and air pollution governance transition (Chou 2015b). And the motivations driving this triple helix have in recent years come from the country's anti-fossil fuel movement as well as the 311 Fukushima nuclear disaster, while considering it from a further distance, we can see that the anti-nuclear and environmental movements have brought greater meaning to the concept of environmental democracy; not only challenging the government's long-term energy subsidies and economic development model, but at the same time in response to demands from civil society we can see an attempt to transform from a top down technocratic model of policymaking to a picture of energy democracy.

The main purpose of this paper is in tracing back roots in Taiwan's environmental movement to understand the multi-faceted elements affecting its development and the way they have influenced Taiwan's current attempts at energy transition; while at the same time revealing the endurance of such social transitions: the historical conflicts between the technocratic government policymaking methods currently in place and the push towards social participation within the policymaking process; finally showing that this energy transitioning has already gradually developed from a civil society movement to a middle class movement, with a turning point in sight.

Reflexive governance and social robustness

Under a structure of both vertical pressure from international green conventions and horizontal pressures from domestic environmental movements, the energy transition movement touches on a series of complicated policies, regulations, social values, choices, and political conciliations, which means that in researching this issue, there is a need within this mosaic of local society policymaking model and regulatory structure to first fit the pieces together, before developing an understanding of this critical reflexive governance, exploring the relationship between the government, industry, markets, society, and scientific assessment (Grin 2005; Beck 2006; Hendriks and Grin 2007; Voß *et al.* 2009). As we see, on the one hand this involves risk policy models, while on the other touching on socio-technical imaginaries, especially with regards to analysis on the unique qualities and throwback residue existing within social development with countries in the stage of development, an understanding of which is vital. Take Taiwan and East Asian countries as an example, developmental states give greater power to the political elites' technocratic form of governance and expert politics policymaking model, moreover these countries so called brown economy is solely driven by a push to see a growth in the GDP in an attempt to preserve the cornerstone of legitimacy to rule. As a result, both fossil fuels and nuclear power are currently the vehicle driving energy governance, and this needs our continued study.[1]

Within social transition theory, researchers' focus is on reconstructing and constructing the complicated technological and social interweaving systems of health, ecology, agriculture, energy, economy, and income disparities as they are involved in the social transitioning process; encouraging local knowledge and the involvement of participatory knowledge in social value choices (Grin 2005; Voß et al. 2009). And this perspective is also useful in the observation of energy transition society, that is to say, we need to discuss the whole concept of energy policymaking, socio-technological imagination, social choice, and other such historical and structural elements from the analysis approach of reflexive governance.

Beck (2010) believes that the economic growth and environmental controls shaped by East Asian authoritative governments' technocratic elite pandering to industries, in turn leads to a unique green Asian capitalism (Wong 2012). Of which in part, economic productivity and environmental regulations become the scapegoats to national authoritative gain; where an ideology which promotes complete dependency on economic development, the market economy not only lacks green thinking, but also vital monitoring from civil society, which produces a predicament in the transition process. Such an observation on the surface appears to present the unique development characteristics of East Asia's early development years, however, it fails to fully grasp social pulsations of the newest developments taking place as this region's society moves from a once undemocratic society up until the late 1980s to seeing in recent years the emergence of a civil society.

In reality, East Asia's newly emerging industrialized countries having been through a crystallized, compressed version of industrialization and democratization, formed technocratic expert politics led by technological governance models (Chou 2007a, 2007b), and as a result of policies whose logic promoted economic development above all else clearly formed a high-carbon regime which deprived these nations of environmental, labour, health, and human rights policies beneficial to the citizens of these countries. However, on the other hand, this expert politics led form of high carbon regime has over the past decade or so come up against increasingly severe challenges from both civil society and public knowledge in the areas of social development, risk regulation, and environmental protection. Civil society increasingly has the capability to act as a watchdog or even flip government policy. Here I would borrow the concept 'transformative capacity' raised by Weiss (1998) who proposed that by observing developmental states economic guided policies' role transformation, one could analyse the governance relationship between civil society and government on issues such as climate change, technological risk, and even energy transition. One would see, namely, whether or not civil society had enough knowledge and mobilization capacity to act as a watchdog to the government; while also observing the way in which the latter responded to increasingly robust civil demands, implementing democratic participation risk policy. Before entering into a structural analysis of the early stages of Taiwan's energy transition history, we should first investigate the vital technological democracy's social robust structure.

Social robust knowledge and civic epistemology

Nowotny *et al.* (2001) advocated that when faced with the many varied threats derived from modern technology, in order to seek for a sustainable existence, society must persistently provide robust responses, reflecting the challenges that such technology poses for the environment, ethics, health, and politics. Most importantly here is that within the knowledge production process, there would be a paradigm shift, that society would not only act as a passive receiver of technological knowledge but rather would change to become active partners participating in increasing one's knowledge; as a result of which social knowledge would be able to provide checks in the expansion and revisions of technology.[2] The unitary method of only recognizing technological experts as reliable knowledge in the foundation building level of government policymaking has already become redundant in the face of a complicated web of climate change and technological risks; such a method is unable to ensure a complete technological explanation and guarantee, calling into question the credibility of government policies by the public (Weingart 2008; Miller 2008).

And with the ability to proactively produce knowledge, society will be able to strengthen its monitoring of government policies, in turn having a greater impact on the environment and health and challenging government and industries wrongful acts (Nowotny *et al.* 2001; Nowotny 2003; Jasanoff 2003; Stirling 2007; Delvenne 2010), in turn confronting risk society's complicated manipulation and ignorance in the relationship between technological knowledge and expert politics (Grin 2005; Rip 2010a, 2010b).[3] The participatory knowledge produced by citizens has the potential of breaking the monopoly of expert politics and enabling the practical application of technological democracy (Jasanoff 2005).

The knowledge productivity, expert role, knowledge competitiveness, and civic epistemology proposed by Jasanoff highlighted by this type of concept of social robustness are relatively consistent. Jasanoff (2005) discusses the democratic movement and foundation of technological policymaking from the perspective of a western constitutionalism form of government, points out that only through systematic programmes can the production, effectiveness, synthesizing, circulation, application and giving meaning to risk knowledge will be ensured. That is to say, policymaking risk knowledge within democratic procedures and systems is in a constant state of competition, in order to gain policymaking legitimacy (Miller 2008, 2015).

Within these competing interests, there is a high degree of competition in the shaping and intersection of knowledge itself. Different actors compete on the stage of technological policymaking in the realms of knowledge and ethics, including the cognitive acknowledgement and explanation of domestic and global problems, as a result of which forms of epistemic politics are developed (Miller 2008). To extend on this idea, we can see that different actors depending on their position and network will form different epistemic communities and through various paths develop epistemic authority, which in turn compete. A democratic form of government ensures that each actor's knowledge and values are guaranteed to enjoy the social and institutional protection to free speech and competitive ideas in order to become an important mechanism in resolving current technological disputes.

However, the concept of an institutional knowledge competitiveness existing within a democracy, debates on policymaking ethics, public participation, and other concepts emphasized in civic epistemology, once transplanted into an East Asian context, would come up against many challenges. The regulatory mode and culture emphasized in the aforementioned analysis of Western countries can't be used as an umbrella to explain the policymaking culture formed by long-term technocratic political history and background present in East Asian cultures; something I refer to as 'delayed hidden risk culture' (Chou 2000, 2002, 2007a,

2007b, 2015a, 2015b). The latter not only refers to the monopoly which the technocratic expert politics have in technological policymaking, but also includes other structural delayed developments and problems in civil society, media, and public awakening. That is to say that no matter whether it be in regards to the way that the regulatory state and society form risk dispute or government policy theory, overall there are significant differences between the East Asian and Western societies' risk culture, and these differences have a direct influence on the politics of policymaking and regulatory culture.

In terms of the issue of energy transition, such a structure of observation is quite applicable. The public knowledge of energy produces a fundamental challenge to the policymaking legitimacy of the current energy regime and this process in itself is a form of 'knowing doing', through the development and construction of risk and energy knowledge, they come to challenge and even overthrow government policies, while through the process of mobilization, civil society is in a continuous process of learning which in turn expands the robustness of public knowledge. From Stirling's (2014) perspective, this is a type of 'Action as knowing' development process. And in particular need of contention is the fact that civil society and rulers have different socio-technical imaginaries, and in accordance with this, they advocate different social-technical systems (Urry 2011; Miller 2015), or, through this development, different socio-technical regimes (Stirling 2014), for example, in relation to the nuclear imaginaries, risk and policymaking. Or the fact that behind support for fossil fuel, there is a complicated system of politics, as well as social and economic systems (Miller 2015) known as high carbon capitalism (Urry 2011).

This can especially be seen in the current energy regime's complicated mechanisms for regulatory arbitrary knowledge, power and economic benefits and the asymmetrically structured agency (Stirling 2014) of which it often consists, making the public challenge energy policymaking and suggesting that energy democracy will not be such smooth sailing. And the experience of western countries in the development of energy transitioning, transferred to an East Asian technocratic expert politics led policymaking model, will be even more of a bumpy ride. Even when considering that during the past few environmental movements in Taiwan we have observed Taiwan's risk policymaking has already gradually began to move from the expert politics path of the past towards a social participation path (Chou 2013a, 2015a), yet this process still needs to be systematically observed.

In fact, this paper advocates that with this energy transition process, climate change and reducing of carbon emissions is the first helix in the chain; on the issue of energy fuel options and carbon emissions there is a significant difference between the government and civil society in the way that this issue is perceived, in turn affecting industrial transformation and air pollution, the other two helixes on the chain. As Jasanoff (2010) points out, global climate change knowledge must be transformed from the larger scale of scientific knowledge to the smaller scale of meaning, implementing social action to influence and explain concepts to both groups and individuals. Climate change, at the level of social action and values, due to different stakes, people have different morals and epistemology, different understanding, as well as social and ethical standpoints. Within the triple helix, we can indeed see the government, industry, related social groups, or civil society all possessing different risk assessments on a variety of issues under the banner of climate change, such as fossil fuels, industrial transition, and PM2.5 air pollution; advocating different knowledge, values, and even ethics. Therefore, within a risk society full of technological and moral disputes, we need the perspective of reflexive governance through which to view the energy transition movement within its historical and social context.

Returning to the capability perspective of developmental states in treating the issues of climate change and energy governance, the first thing we need to grasp is whether or not Taiwan possesses social robustness, namely, whether civil society is robust enough to systematically produce risk knowledge, participate and monitor government policies as well as proposing different technical or embedded value. And this transitional capacity is vital to an understanding of government and civil society and the complicated democratic governance of climate change. The more the two sides cooperate, the stronger the transitional capacity; while the greater the stand off between the two, the greater the difficulty on the road to transition. From an initial view, the triple helix of Taiwan's energy transition movement, are all driven by civil society, no matter it be fossil fuels and carbon emissions, industrial transition, or air pollution governance issues; civil groups have been gradually able to systematically build risk knowledge, with a robustness that has enabled them to challenge the governmental policies and bring about public awareness. And this process requires us to study these developments in the context of specific historic events.

Authoritative expert politics and nuclear regime

The process of energy transition in Taiwan has already been underway for more than 30 years, as far back as 1985 with the dispute over new plans to build a fourth nuclear power plant in Taiwan, then with the Chernobyl nuclear disaster in Ukraine in 1986 this dispute intensified, growing into the anti-nuclear movement. In recent years the extent of Taiwan's carbon emissions and pollution gave rise to the anti-Kuo-Kuang Petrochemical movement in 2010 and then in 2011 the 311 Fukushima nuclear disaster reignited the intensity of civil society protests against the fourth nuclear power plant, which led later in 2014 to the successful suspension of its commercial operation. To understand these developments, we need to return to the authoritative political historical background of the cold war and the technocrats and technocratic elite led technocratic expert form of governance. A form of governance which combined a history which was at once driven by both military considerations and a push towards economic development, on the one hand forming a dictatorship of a nuclear power regime, while one the other hand in a political climate that had not yet begun the process of democratization, the technocratic policymaking culture has continued well on through till today, resulting in the development of a delayed hidden regulatory culture.

While after the Second World War, the use of nuclear power had already transitioned from a military to peace time form, in the 1960s, under the framework of a global cold war, the Chiang Kai-Shek regime continued in its attempts to build nuclear weapons in order to intimidate and respond to the threats from Communist China. At this time the US also exported large numbers of nuclear power plants to Asia; by the end of the 1960s Taiwan had already begun its plans for building its first nuclear power plant and by November 1970 they had already started building it. By July 1984 Taiwan's first nuclear generating unit of a third nuclear plant was already under commercial operation, forming a complex relationship between the military and the nuclear industry. Even though the state-run Taiwan Power company (hereafter TaiPower) proposed plans to build a fourth nuclear power plant in May 1980, contending political, economic benefits and rocketing construction costs meant that on 27 March 1985 55 ruling party congressmen made a petition for the suspension of the new nuclear construction project; the following year with the Chernobyl nuclear disaster occurring on 26 April 1986, Taiwan's anti-nuclear movement was born.

The anti-nuclear movement was not only about social protests; on 16 July 1987 Taiwan saw an end to martial law and the process of democratization began, as a result of which there was an outburst of political movements including protests against anti-pollution, and labour or student-led protests rose up like bamboo shoots after the rains (Hsiao 1987, 1993); and the anti-nuclear movement was no different, deconstructing the technocratic regime. 'Anti-nuclear protests are protests against dictatorship' (Lin 1989); 'the issue being challenged in the anti-nuclear movement, besides the pros and cons of nuclear power and the risk assessment, there is also the struggle over policymaking legitimacy and process, such as the phenomenon of the technological and central government dictatorship' (Shi 2002), 'opposing technocratic hegemony and media monopoly, the language of the anti-nuclear protests is one against the dominating patriarchal mentality' (statement by the National Female University Students Active alliance against 4th Nuclear Plant) (Shi 2002). These show that Taiwan's energy transition movement began as a protest against authoritative government, technological hegemony, gender dominance, and other issues in the emerging atmosphere of developing democracy and within this, intellectuals played a key role in producing risk knowledge and mobilizing the public at a local level, becoming a vital element in developing social robustness. In 1987 a number of university professors formed the Taiwan Environmental Protection Union, bringing together the local public in the designated area for the fourth nuclear power plant in Taipei county's Gong-Liao village, to fight against the plan, gradually expanding the base of the anti-nuclear movement. At the beginning of 1988, anti-nuclear academics joined together to invite environmental protection, political, religious, women's, and aboriginal groups to release an 'anti-nuclear declaration', and in the same year they also invited West German Green party members to support Taiwan's anti-nuclear movement.

With this process of mobilization, the anti-nuclear movement nationally reached its peak in the early 1990s: in 1994 Gong-Liao village held a village referendum and showed that 96% of residents were against the building of the fourth nuclear power plant, and nationally marches against nuclear power were in the tens of hundreds. Despite all of this, the process of democratization was unable to reverse the technocratic expert governance style dictating policymaking. During this time, we saw university professors leading civil groups, strategically building alliances and mobilizing local groups from different fields, developing a strong discourse of risk through marches, lectures, and other methods. However, from my perspective, while this was the first time in Taiwan's history that civil society had emerged with a relatively systematic form of social robust knowledge, yet at this time they were unable to break through the mix of technological elitism and technocrats at work in the nuclear energy regime, a legacy of decades of cold war authoritative governance, and the later, by displaying the vestigial residue of an authoritative regime culture, continued with the dominance of a scientific rational expert politics. Moreover, despite the intense process of democratization in the 1990s, this authoritative expert politics and regulatory culture has continued to persist until today, increasingly coming into direct confrontation with an increasingly robust civil society (Chou 2007b; 2009).

Nuclear complex in Taiwan

The nuclear energy regime in Taiwan is similar to that in Japan, combining technocrats, industry, and academics, and even media to form a gigantic nuclear complex. Much research points out that Japan's energy policymaking takes place within a backdrop of a nuclear village formed by the government, the nuclear energy industry, and academics (Sugiman 2014; Aldrich 2013; Kingston 2013) otherwise known as a three way nuclear complex (Matsumoto 2013) which controls the whole process; and this dominant network not only ensures the power companies' massive financial gain, but also manipulates related data, displaying the extent of this complex's political power (Funabashi 2012: 68). Such a policymaking structure, is what we ordinarily see, under the veil of technology, experts, and economic development, lacking in risk communication with the public, and even manipulating and hiding information. And a similar story has been seen in Taiwan, with the Kuomintang party (KMT) having been a long-term supporter of nuclear power, no matter it be in 2000, with the party being in opposition for the first time in its history, or 2008 when it returned to power, both within the party and in its dealings with other outside parties, the party has sustained its firm technocrat-TaiPower-Academic nuclear complex, dominating the energy policymaking process. Even when in 2001 the Democratic Progressive Party (DPP) was in power, the Executive Yuan passed a 'nuclear free homeland', and proposed a bill to put a stop to the policy pushing for the building of the fourth nuclear power plant, the large number of KMT members in parliament meant that the KMT were not only able to successfully put a stop to the DPP's proposed bill, but through the political storm brought about by this political manoeuvre, they attempted to impeach the then president, Chen Shui-Bian.

The development of renewable energy has come up against the same contradictory structure. In 2005 when the then ruling DPP proposed Taiwan's first renewable energy draft law, it was immediately blocked. Even in 2008 when the KMT came back to power, and in view of international trends and President Ma's personal attention, in June 2009 the Renewable Energy Development Act ratified, however under the domination of the nuclear complex, the development of Taiwan's wind energy and solar energy have both been comparatively slow. Other than facing bureaucratic interference in the application process, there is also opposition from technocrats, technology related academics, TaiPower, and the TaiPower Union who strongly support nuclear power, constructing a structural path dependency: cognition of policy-making, bureaucratic technical interference, nuclear power technical imagination, domination of social discourse, all stand in the way of renewable energy's development. For example, in view of international trends towards reducing carbon emissions, in 2010 the Executive Yuan proposed the 'Energy Saving Carbon Reducing Action Plan', however, within this policy nuclear energy was seen as an important 'two high two low' low carbon energy tool. Under this kind of development, Taiwan's renewable energy accounted for less than 1% of all national power between 2001 to 2014 (Chou 2015c).

This technocrat-TaiPower-academic nuclear complex has formed a normal system of operation. For example beginning in 2008 the committee chairman and vice chairmen of the Atomic Energy Council of the Executive Yuan in charge of Taiwan's nuclear energy policy are both graduates of Qinghua University's department of Nuclear Engineering. Moreover committee chairman Tsai Chun-Hong failed to avoid conflicted interests when he also served as the director of the Nuclear Society, where in the aftermath of the 311 Fukushima disaster he advocated in the Nuclear Society's periodical that 'Taiwan's nuclear plant is an earthquake liberator', and that 'the designs of our nuclear power plants are not as sturdy as a rock, because the plant is itself a rock' (Yang 2011), and this in response to the Vice Chairperson Huang Qing-Dong's statement on the 311 disaster 'Nuclear Power Plant 4 will be built upon the rock, as safe and sturdy as buddha sitting upon a lotus pad' (Liu and Tang 2011). Such comments were responded to with strong opposition from environmentalist groups, moreover at the beginning of February 2015, former director of the Environmental Protection Agency Chen Shi-Hong who had resigned from the position no more than one year previously, was appointed as head of Sen-Ba Power Plant, a company which was heavily invested in by TaiPower. Besides being fairly suspect in violating the environmental assessment procedure currently being carried out at the plant, the EPA is also responsible for environmental and health controls for Taiwan's nuclear power radiation levels and nuclear waste. Coincidentally, even though on the 20 May 2016 the KMT once again lost power, within ten days of the newly established DPP led government taking power there was a power shortage threat by TaiPower, which ignited fear in the general public. For an individual to serve as Chen Li-Cheng has to be president of a company like the GIBSIN Engineers, Ltd. responsible for assessing the electricity capacity and as a peripheral policymaking organization of EPA by the KMT regime, as well as having served in the role of consultant to various departments' energy policymaking bodies as a professor at Qinghua University, using their positions to persistently advocate for a renewal of plans to bring the 4th Nuclear Power Plant into operation, while belittling the efficiency and long term potential of renewable energy sources (Chen 2016; Ye 2016; Lei 2016) is a completely inappropriate misuse of position.

The analysis in the previous discussion shows that despite seeing changes in political parties more than once during the past decade or so, the nuclear complex which exists within Taiwan's energy related policymaking bodies remains powerful, and continues to affect Taiwan's energy transition process; while with the current failure to liberalize the power industry, TaiPower continues to monopolize the generation, transmission, distribution, and sale of electricity, moreover they maintain a relatively strong grip on all various manners of energy related data. To put it another way, there currently exists an intertwined nuclear complex network which continues to restrict the recently elected DPP government in what they can achieve. And one area for contention is the control of power related information at a national level with the concept of public communication.

Explaining these developments from the perspective of risk culture, one could say that East Asian countries, including Taiwan, have historically, due to their late-coming, catch up role and therefore a desire to accelerate or compress the time for the industrialization process, civil society has in the past been unable to accumulate a strong capacity to provide checks and balances. Therefore a weak civil society, in the face of a rough monopoly of technology economy, has led to the forming of systematic hidden risk in a delayed high-tech risk society (Chou 2000; 2002) as well as other compressed modernization problems (Chang, 2010). And the nuclear complex which has developed in both Japan and Taiwan is fundamentally a product of such historical conditions; where even changes in political party or even the form of government and development of civil society, have failed to rapidly replace or flip the impact of such a complex's existence on energy policy.

Take Japan for example, where this complicated complex driving modern nuclear energy policymaking, where key players including politicians, scientists, engineers, public relation experts believe deeply in technological development and with the so called safety myth bringing about a global trend towards a renaissance in nuclear power, a new form of social risk discourse has emerged (Aldrich 2013: 250). And this has also occurred in Taiwan. The movement against the fourth Nuclear Plant reached its peak in the 1990s, however due to the departure of the DPP from power in 2000 there was a failure to terminate the construction plans for the fourth nuclear power plant, and this, along with TaiPower and the nuclear complex persistently expressing that without the 4th Nuclear Plant Taiwan will lack power has drawn the attention of the public, leading to the anti-nuclear movement losing its mobilization capacity and facing a serious defeat. At the same time, the issue of global warming led to a global renaissance in nuclear power which also affected Taiwan, and with the dominance of the nuclear complex's discourse on nuclear safety and fear of lack of power, people began to believe that nuclear power could take the place of thermally generated power, becoming an important tool in the goal to reduce carbon emissions. In 2008, Taiwan's highest academic research body, Academic Sinica released a proposal paper 'Energy Policy in responding to global warming', which directly recommended that 'the 4th Nuclear Plant should be completed according to original plans, while Nuclear plants 1, 2 and 3's lifespan should be extended to substitute traditional power, and consideration should be made for adding new nuclear units at already existing power plants' (Academia Sinica Environment and Energy Research Group 2008). Such a mythology of nuclear discourse continued until it was destroyed by the reality of Japan's 311 Fukushima disaster, when the public returned to their anti-nuclear conscious.

Triple helix movement of energy transition by social robustness

In the past 20 years the process of energy transition in Taiwan has been taking place under direct pressure from international climate change treaties as well as domestic pressure from the environmental protection movement, and in recent years has received growing attention, being considered an increasingly important issue in the face of the 311 Fukushima Disaster of 2011. In 2010 the anti-Kuo-Kuang Petrochemical construction movement which had begun in the 1990s in opposition to high levels of pollution and high energy consumption, brought Taiwan's environmental movement from a protest against pollution levels to a movement which focuses on the risk of climate change and the need to reduce carbon emissions, producing momentum for an industrial transition (Chou 2015b); at the same time, the anti-Kuo-Kuang movement and the PM2.5 air pollution issue dispute, became the turning point for developing an anti-air pollution movement. And the direction of these two developments came together in the aftermath of the Fukushima accident to propel forward the issue of nuclear risk. Beginning in 2011, these issues evolved into a mutually agitating influence component, bringing an outright assault on energy policy, a process which I refer to as the triple helix movement.

First spiral: industrial transition

The petrochemical industry has played an important support role to this newly emerging industrial country's manufacturing industry; beginning in the 1980s civil investment has seen this industry's continued expansion. And since the 1980s the petrochemical industry has provided an important basis to Taiwan's economic growth, with an annual value of production and staff numbers which accounted for a third of the total for the manufacturing industry, with an export value which also accounted for a third of the national total, becoming the largest manufacturing system in Taiwan (Tsai 1997). In 1992 the Sixth Naphtha Cracking Plant was established, with the fifth Naphtha coming into operation in 1994. And by 1998 as the sixth Naphtha Plant came into operation the national level of carbon dioxide (CO_2) emissions reached an all time peak; between 1992 and 2000, this industry became the culprit for levels of CO_2 emissions increasing by 1.3 fold (Chou & Liou 2012).

In 1993 when the Taiwan government began planning the Bin-Nan Industrial Complex, local and national environmental groups rose up in protest, and after more than a decade of conflict finally saw the end of the planned project; during this period, while in view of pressure globally from climate change, the government in 1998 responded to the Kyoto Protocol with the first national energy conference. In 2001 they established the National Council for Sustainable Development (NCSD). In 2005 the second National Energy Conference was held, while in 2006 meetings began to take place of the NCSD, with discussions taking place on many issues related to the brown economy. However, within the cabinet there remained plans to attempt another proposal and in 2010 the proposed expansion was officially announced to international outrage. Since the anti-Bin-Nan movement began in 1995, civil environmental groups' protests have been gradually developing from local anti-industrial pollution to protests on a national level against high energy consumption, high CO_2 emissions and high pollution levels, or even high water consumption becoming a more complicated movement in response to climate change risk as a result of which the protests against Kuo-Kuang Petrochemical project reached its peak.

From protests against Bin-Nan petrochemical plant and steel movement, to the later protests against Kuo-Kuang Petrochemical construction, we can see a progression as the level of protest was raised from a focus on pollution to a higher focus on climate change risk. Most importantly though is that this mobilized the whole of society to reevaluate these issues in a very real way, critically considering the idea of a brown economy and attempting to construct a concrete image of a green economy to drive the push towards industrial transition which involved a move towards the discarding industries with high levels of energy consumption, pollution, and water consumption as the discourse on the risks of climate change continued to develop. There was, one could say, a reflective constructing of Taiwan's energy and economy towards sustainability, molding a new social development framework.

In particular, in recent years the movement against Kuo-Kuang has seen the bringing together of various different civil groups, including the Anti-Kuo-Kuang Petrochemical Discourse Alliance made up of university professors, Changhua Environmental Protection Alliance, Youth Students Anti-Kuo-Kuang Alliance, National Arts Group, Changhua Medical Alliance etc. to form a strong body, systematically developing social robustness. Through producing risk knowledge they are able to continue to challenge and flip the government's development discourse. As part of this, in terms of industrial CO_2 emissions, high energy consumption and contributions to the GDP, wetland ecology, land subsidence, white dolphin conservation, village sustainability, air pollution, robust risk and social economy assessments etc. all work together to bring about a solid risk discourse. *And these encapsulate to form a picture of how Taiwan's society and economy will develop in the future, building the first driving helix of energy transition, namely the direct relationship between brown industries and problems of high energy consumption and CO_2 emissions.*

The Anti-Kuo-Kuang Alliance made up mainly of university professors pointed out that the contribution that the petrochemical industry made to the national

GDP was relatively low, when compared to the huge damage it was having on the country's environment: 'in 2008 Industry accounted for 48% of Taiwan's two hundred and fifty-five million tons of CO_2 emissions and of this the Petrochemical industry was responsible for 35%. Yet the petrochemical industry only accounted for 18.67% of Industry's production value'[4], 'currently the average CO_2 emissions in Taiwan per person is 12.08 tons a year, a figure more than twice the world average of 4.38 tons per person per year, higher than the OECD countries 10.97 tons per person per year average, higher than Taiwan's neighbouring competitor countries of Japan at 9.68 tons per person per year, South Korea 10.09 tons, per person per year ... if we estimate that Kuo-Kuang is responsible for emissions of 12 million a year, that would mean an increase of 0.55 tons per person per year for every individual in Taiwan'[5].

While in terms of conflicts over water resources, the Anti-Discourse Alliance criticizes Kuo-Kuang for using more water than Changhua county's water reserves can handle: in one day Kuo-Kuang will use an estimated 400,000 tons of water, which is more than 366,000 tons used daily by all the residents of Changhua put together. And with the lack of balance in the water resources supply and demand, the result is one of increasingly severe land subsidence (Chou *et al.* 2010).

In particular, this alliance points out that whenever Kuo-Kuang Petrochemical company builds a new plant PM2.5 levels rise drastically, spreading in high concentrations, moreover the alliance calculates that 'the numbers of individuals entering hospitals and dying due to respiratory tract, blood vessels related diseases increased to 339–515 people per year, while at the same time causing the average person to lose 23 days from the average lifespan' (Anti-Discourse Alliance). Besides this, it also influences agricultural food safety: 'if making a comparison of levels of air pollution emissions produced by Kuo-Kuang and Formosa Plastics Corp's (FPC) sixth naphtha cracker complex, ... the PAH, heavy metals, dioxin and other organic carbon compounds are all Carcinogen, after being emitted from industrial plants the polluted emissions fill the air, water, land and settle in agricultural produce, affecting the food chain and mother's milk through produce such as milk, duck eggs, vegetables, oysters' (Chou *et al.* 2010).

This systematic industrial and energy discourse, also formed a wide social development framing, bringing about a conflict of awareness between environmental concerns and economic development, becoming the first instance of public consciousness being flipped by awareness of the need for great environmental protection in the process of economic development. According to the *China Times Daily*, in April 2011 the government decided to put a stop to large-scale national development as it found that national opinion was strongly against the Kuo-Kuang development with those against much higher than supporters. This survey's results demonstrated a break with the past when the Taiwanese public, while possessing a general understanding and support for environmental protection, would still support large-scale development projects because of the desire to see the country's economy and development continue to flourish.[6]

In other words, the professor-led anti-Kuokuang Discourse Alliance, through constructing robust risk knowledge, were able to successfully break the monopoly of technocratic expert politics, proposing the competing lines of knowledge and morals, to create new forms of social participation and methods. This, in turn, led the Executive Yuan to establish a board specifically for exploring this issue, while after the Rio+20 Conference, a green economy working small group was successfully established (Liu 2012), re-examining Taiwan's brown economy and industrial transition.

Second spiral: anti-PM2.5 movement and air pollution governance

Anti-Kuo-Kuang Petrochemical project also extended people's awareness of PM2.5 airborne particles. Due to the transition process of Taiwan's industrial structure having been delayed, levels of energy consumption, pollution, etc. meant that while the environmental protection agency had early on researched and proposed methods for reducing PM2.5 airborne particles to keep in line with the air quality standard, in reality levels in the air are currently high and only controls on overall total suspended particles (TSP) are maintained, that is only levels of PM10 and more are regulated. And such a serious health risk issue was only brought into the limelight with the protests against Kuo-Kuang, when discussing the organic particles produced in the petrochemical manufacturing process. In fact, before this, there had already been large amounts of research performed which highlighted the fact that the petrochemical industry was producing air pollution which posed serious health risks to the general public (Hsu et al. 2003; Chan 2012; Tsuang and Kuo 2011).

Moreover, just as the European Environment Agency (2001) has criticized, the delayed hidden nature of these controls caused an institutional ignorance, which over a long period of time also led to societal ignorance. What is even more serious is that retroactive thinking will continue to delay the transition process in Taiwan's brown industries, while the high energy consumption, pollution, etc. caused by the delays in the industrial transition process has directly affected the supply and demand aspect of energy transition. And once these different aspects have formed a negative cycle, the only way of resolving the lack in supply and demand is to pursue nuclear energy and thermal power generation to supply the increasing demand for those high consumption industries. That is to say that brown industries are themselves dependent on large amounts of energy supplies, thereby creating a demand for high risk nuclear energy generators and thermal power generators with their high pollution emission levels: the deep connection between brown industries also at a deeper level is connected to the persistent use of brown energy.

However, the anti-Kuo-Kuang movement has not yet fully removed the mask from this problem, but rather taken the first step in revealing surface level pollution; and while this environmental movement has already reconsidered the problems of CO_2 emissions and high energy consumption, it remains limited in its critique of the petrochemical industry as a whole; as such a reflection on energy risk threats came in the aftermath of the 311 Fukushima accident.

In the past the professors' assessments of PM2.5's threat to human health became a key issue in the protest movement against Kuo-Kuang, calling medical groups to sit up and pay attention, as well as mobilizing and extending risk awareness as a whole. After 1300 university professors had signed a petition against Kuo-Kuang's planned development, and then held a press conference on 3 August 2010, the medical world followed suit on 6 August in the Legislative Yuan with a press conference. By comparing lung adenocarcinoma cases in different areas of the country they were able to point out that in areas with petrochemical industrial plants, high overall suspended particle emissions increased the public's probability of lung cancer and cardiovascular diseases. This support and strengthening of the professors' analysis meant that PM2.5 as an issue developed to become an important axis in Taiwan's environmental movement.[7]

Besides attending the 'Petrochemical Policy needs to make a Swerve, Environmental Protection Save the Country March' organized by the Changhua County's Environmental Protection Alliance on 11 November 2010, the Changhua medical community and more specifically many experts in the field held a protest outside the Nanyao temple on 6 December 2010. They protested against the building of a new plant in Changhua on the grounds that the petrochemical produced pollution causing lung cancer and calling for the local government to immediately put a halt to the bid to build a plant in Changhua (Tsai 2010). At the same time, in order to further highlight the health risk caused by PM2.5, local doctors established the Medical Alliance for Public Affairs in March 2011, proactively getting involved and making public the effects that the building of a Kuo-Kuang plant in Changhua would have on Central Taiwan, as well as the health threat it would pose to the general Taiwanese population as a whole. They produced a small leaflet named 'PM2.5 and health', and also made use of YouTube to spread more widely risk knowledge on this subject.[8]

Following the announcement by the government on 22 April 2011 that it no longer supported the plans for expanding Kuo-Kuang's petrochemical plants, Changhua's medical community's Medical Alliance changed its role to become an important against-expert group which continuously played the criticism on the air pollution governance. That is to say that, those involved in the anti-Kuo-Kuang movement were in particular able to raise awareness of the threats to health that come from PM2.5 airborne particles, and even more specifically, as with the university professors, the medical community group were able to develop and strengthen social robustness using statistics to produce a health risk discourse, while having an important effect on policy.

In the first stage, this independent medical expert group was mainly able to convince people of the real risk and threats that petrochemical pollution can have on people's health and by doing this they were able to see the development of strong civic knowledge and social robustness; while in the second stage the issue of PM2.5 was extended to consider the effects of not only the petrochemical industry but also the steel and coal fuel, revealing a more diversified group able to protest against this issue. *That shaped the beginning of the second helix movement in Taiwan's energy transition.*

In fact, following the halting of Kuo-Kuang's plans, and in the aftermath of the 311 Fukushima disaster, people began to focus on energy security, with the whole of society rethinking sustainable industries and clean energy (Chou 2013b). Therefore, not only did the risks of nuclear power become a topic for debate, but also the high CO_2 emissions and airborne pollution caused by thermal power became a hot issue. Increasing numbers of the general public believed that Taiwan should follow in the footsteps of industrialized countries, making an effort to develop renewable energy on a larger scale, in order to avoid future nuclear energy disasters or the health risks associated with airborne pollution.

The issue of PM2.5 expanded and has taken a new turn, with the Changhua medical community alliance taking the first shot. On 24 April 2011 the alliance member Doctor Lin Shi-Xian openly criticized the government projects which plan to build river weirs on the River Dadu and the adding of components to the existing Changhua coast thermal power plant, causing severe drying out on the downstream riverbed, in turn producing severe sandstorms and increasing the spread of coal fuel air pollution (Lin 2011). In May of that year one member of the group, Doctor Ye Guang-Fan, organized 'Taiwan healthy air working group'; on 5 May they held a special Mothers Day Press conference, demanding that the government begin to regulate PM2.5. As a result of such pressure, on 14 May 2012 the environmental protection agency (EPA 2012) officially announced plans to regulate PM2.5 levels.

Under this context, there emerged a variety of anti-air pollution activity and new groups, gradually demonstrating the robust capacity of Taiwan's civil society. Since September 2012 a group monitoring the 6th Naptha Plant in Yunlin County's Mailiao has been releasing the 'Since 6th Naptha Came' which brings together criticisms of the pollution and coal fuel pollution it produces. In 2012 the EPA of US (2012) increased the seriousness of PM2.5 annual regulatory standard, from 15 microgram to 12 and this intensified environmental groups demands on the government. Then in November 2014, the long-term Kaohsiung-based Citizens of the Earth group, monitoring petrochemical pollution in the city, began a movement encouraging Kaohsiung's elementary school kids to hand out air pollution flags, the result of which had national implications and response. At the end of 2014 Nantou County's Puli, affected long-term by the trans-boundary impact of both Yunlin County's 6th Naptha Plant and TaiChung's Thermal Power Plant, saw university professors, students, and local residents join together to form the 'Reduce Puli's PM2.5 Air Pollution SOS' gradually developing citizen science with voluntary monitoring of areas to measure air quality (Tu 2016). In March 2015 a group of young directors established the 'Kaohsiong Good Air SOS group', planning to film a series of short pieces through which they hoped to influence government policy. On 8 May 2015 on the eve of Mothers Day, the Homemakers Union for Environmental Protection formed to ensure healthy mothers and mother-in-laws. They held a press meeting in Taichung demanding that various counties and cities as well as central government should care more about healthy air in order to prevent a serious health issue for the next generation and proposing concrete policies. Due to its touching on issues of the next generation's health, this has gradually shown signs of becoming a middle-class movement.

At the end of February 2015, the self-financed documentary *Under the Dome* of former China Central Television reporter Chai Jing on China's air pollution, brought all eyes internationally to fix on China's current air pollution issues, including a critical narrative of petrochemical, steel, and coal plants causing serious PM2.5 pollution and increasing Taiwan society's demands for sustainable industries and clean energy social framing. The robust proposals from different environmental groups for increasing risk knowledge and discourse directly influenced government controls and regulations. On 4 April 2015 Yunlin County government launched a joint signed oath that called for central and southern Taiwan cities and counties to sign a shared vow that they would no longer burn raw coal and petroleum coke. Taichung, Changhua, Yunlin, Jiayi and Tainan altogether five counties together signed and supported this oath. On 15 May at a Yunlin County meeting 'the industrial business plant is forbidden to use raw coal or petroleum coke self governance legal regulation', *after this the Taichung government also proposed, 'the Public and Private Control Regulations for Raw Coal and prohibited use of Petroleum Coke Self Regulation.'*

These developments show that the Anti-Kuo-Kuang movement has developed and spread to become an Anti-PM2.5 movement; from a single group it has developed to become a national and local, diverse and multivariate activity and organization. Moreover, the helix of criticism of petrochemical-produced pollution has developed from a single issue criticism, to become a wider criticism of steel, petrochemical, and coal plants. In particular, coal fuel, and the Taichung thermal power plant with the most components globally as well as various private cogeneration plants, became the focus of civil and environmental groups. Namely, the helix of the anti-air pollution movement gradually changed to focus on the banning of raw coal burning as well as a dispute for choosing which form of energy should be adopted. This unique development differs from other countries in their energy capacity as well as energy transition experiences; currently most countries have shown their concern for nuclear risk and gradually strengthened the development of renewable energy. *However, the complex development of the anti-air pollution movement has already become the second helix and this works together with the anti-nuclear movement, integrating the two, on the one hand demanding the use of non-nuclear power and on the other demanding an end to coal.* And in the promotion of this second helix it has been gradually bound together with the first helix to create an important driving force, which, one can see, has enabled the development of a pluralistic robust civil society in Taiwan. And this kind of development has also been seen with the third helix movement, with a third strong leader in propelling environmental groups.

Third spiral: energy fuel dispute and carbon emission

Before the Fukushima disaster, while the push for energy transition had been going on for more than two decades, the focus had mainly been on the protests against the building of a fourth nuclear power plant. The movement was mainly led by Taiwan's Environmental Protection Alliance, Green Consumers

Foundation, Gongliao residents, LanXia Anti-Nuclear waste aboriginals, and other groups. The anti-nuclear movement began in the 1990s with forty to fifty thousand people taking to the streets to protest against a further nuclear power plant. Yet following this high point interest dwindled and numbers dropped, to the extent that by the mid 2000s the pressures of global warming and the need to drastically reduce CO_2 emissions meant there was even something of a nuclear renaissance in Taiwan. And this can be seen most clearly with the 'Energy Policy Recommendation paper' released by Academica Sinica (2006) and the KMT plans for a 'Sustainable Energy Policy Action Plan' in 2010 which listed nuclear power as a tool for ensuring low carbon energy.

In reality, under the dominant force of the nuclear complex discussed earlier with its long term hegemony of nuclear safety discourse, even when the DPP came to power in 2000 and attempted to take over power in the anti-nuclear arena, the joining of different anti-nuclear groups to challenge the image of nuclear power as a safe form of energy, met with difficulties as a technocrats-Taipower-scholars complex continued to hold a tight grip on this policy-making machine. And this development process, as Urry (2011) points out, is taken from the analytical perspective of those on a path of dependency, analysing how larger scale social technical systems form high carbon operational mechanisms, and by the accumulation of expropriation, support high carbon capitalism. In this way nuclear power has held on as Taiwan's central brown economy.

However, in 2011 in the aftermath of the 311 Fukushima nuclear accident, this entire trend was radically turned on its heels. In terms of time, on 22 April the plans for extending the Kuo-Kuang Petrochemical plant were brought to a halt and this decision was directly related to the anti-PM2.5 movement. Even more pertinent was the way that the weak and dejected movement was rekindled,[9], and more importantly, just as in the development of the previous two movements, participants in the movement gradually changed from environmental groups to encompass expert groups, art groups, women's groups, thus gradually gaining momentum and becoming a middle-class movement. *Within such a historical progression, energy transition is not only an issue of anti-nuclear, but rather a complicated change in response to a reflective critical construction of the ways in which Taiwan can move towards a sustainable economy, environment and substitute energy sources in order to remould the socio-technical imagination.*

In autumn 2012 the Green Citizens' Action Alliance launched a national Anti-nuclear flag: 'Anti-Nuclear, no more Fukushima' activity, which saw the development of a new group of social actors. Continuing on 20 January 2013 with the National Nuke-abolishment Platform, on 8 March during Women's Day, art groups, academics and so on established the 'Moms Monitoring Nuclear Power Alliance', seeking the help of reporters and celebrities and utilizing anti-nuclear propaganda, and reporting through the Internet. The life and security concerns projected through the art snowballed, successfully increasing people's knowledge of the risks of nuclear power and encouraging young people and the middle classes to join in. On 9 March more than 220,000 people took to the streets to participate in a nationwide march, with participators reaching an all-time historic high for such a march, and catching the attention of the world's media. On 8 August, Taiwan's fathers day, a group of writers, artists, and academics who are also fathers responded by forming the 'Daddy's Monitoring Nuclear Power Alliance', environmental groups and more than a thousand shops nationwide joined in an 'anti-nuclear drugstore' event, attempting to bring risk awareness to every area of society.

These events forced the ruling KMT nuclear regime to attempt to transform their policy which had previously allowed plans for a fourth nuclear plant to be carried forward. In February 2014 a proposal for a referendum on the fourth power plant attempted to limit the number of people who could pass the basic participation criteria, thus hoping that the outcome would support not bringing the plans to a halt. But such a move received a major political backlash. In order to prevent the government from pursuing such strategies the former chairperson of the DDP Lin Yi-Xiong protested by beginning a hunger strike on 22 April. This move gave rise to significant political pressure and in the end the ruling party was forced abandon their plans for the fourth nuclear power plant.

The halting of the commercial operation of the fourth nuclear plant was not directly linked to the movement against industries with high CO_2 emissions and high air pollution. However there was already a foreshadowing of these three issues. The anti-air pollution movement began to spread around the country. In February 2014 the southern city of Chiayi established a 'PM2.5 SOS meeting', then in early March 'the Tainan air pollution avoidance SOS meeting' was established. On 12 May Taiwan's Healthy Air Alliance came together with Changhua Medical Community Alliance and various environmental groups at the Legislative Yuan to hold a press conference, highlighting that Taiwan's air pollution was greater than the three Asian little tigers. Demands were made for the government to move fast in proposing measures for tackling the problem and establishing research data.

At the time Taiwan semiconductor manufacturing company based in Taichung's Science Park also consumed a lot of electricity, causing serious and extensive deforestation. This also caused a heavy load in terms of higher levels of carbon emissions, as well as air pollution etc. On 6 June 2015 Action Coalition for Healthy Air in Taiwan held a series of anti-air pollution marches in nine different cities around the country. The issue of a simple air health risk moved to a general criticism of PM2.5 as the cause. The use of coal in Taichung's thermal power plant, 6th Naphtha private plant, and the process of emissions from Dragon Steel and the 6th Naptha Plant were held responsible for large-scale pollution.

Thus the focus of society gradually turned from a simple issue to a complicated criticism of energy, air pollution, and industry, including an examination of brown energy and brown industries. But the most important catalyst for the development of a triple helix movement was in response to the fact that beginning in April 2015 local government attempted to bring in measures to put an end to the burning of raw coal. Environmental groups began proposing the use of natural gas to replace coal fuel, and hoped that renewable energy could increase in terms of proportion of overall power sources, becoming an important tool in the transition to a lower risk, lower carbon economy. This new source of energy discourse continued to ferment. Following the COP21 in December 2015, groups were empowered to achieve the expectations of the Paris Treaty stopping the use of fossil fuels, seriously restraining industrial emissions, while speedily developing renewable sources of energy. Moreover this reflects the fact that following the nationwide Anti-Air pollution march on 26 December, the idea of a 'non-nuclear non-coal garden' became the major appeal (Air Clean Taiwan 2015). The combination of anti-nuclear/anti-air pollution was for the first time brought together under one umbrella. Moreover this process wasn't initiated by the anti-nuclear groups but rather it was the anti-air pollution groups who first recognized the possibility of locking down this environmental issue. On the one hand they demanded that the government practically enforce the Nuclear-free Homeland policy. This was passed in 2001. On the other hand, they stressed that emissions were not solely nuclear risks but also issues of the higher risk of air pollution generated through the already existing power plants. At the same time, this also criticized industries with high levels of pollution and carbon emissions.

These developments have brought a fresh boost to Taiwan's energy transition, beginning 20 years ago with the anti-nuclear movement. Twenty years later, changes in energy fuel choices (energy industries' CO_2 emissions, air pollution, and nuclear risk), industrial CO_2 emissions, and energy consumption all give rise to a debate about the power of social choice. Between the three there has formed a mutual impact momentum, with an active movement to press for changes.

In fact, in the aftermath of the 2011 Fukushima accident, the Taiwan public's perception of climate change and energy risk had already undergone a clear change. According to an academic survey in 2012 (Chou 2013b) and a new survey released in 2015 (Chou 2015c), there were signs that Taiwan society's had undergone a paradigm shift. The part of the survey which showed those 'willing to pay' in 2015 had reached an all time high of 85.1%, in 2012 81.6% of those interviewed were willing to abandon lower valued nuclear energy and 'high power value supporting renewable'; only 10.8% (2015), 12.2% (2012) of interviewees expressed they were not willing. As for carbon figures, at the end of 2015 78.7%, in 2012 the figure reached its peak at 80.6% showing that a large number of people were willing to see renewable energy reduce the scale of greenhouse warming, In contrast, in 2015 54.6% and in 2012 58.5% of interviewees demonstrated that they were unwilling to see nuclear power replace thermal power as a means of reducing carbon emissions.

In support of green tax, 64.4% in 2012 and an increase of more than 4% to 68.8% of interviewees by 2015 showed that they were willing to 'accept a higher tax to protect the environment', while 28.3% (2012) and 24. 8% (2015) of interviewees did not support this. On whether people supported the adoption of a government energy tax or environmental tax, we can see that in 2012 61.1% of interviewees supported such a policy, a figure which had increased by 7% by 2015 when it had reached 68.0%. On the other side, the figures for those against such a tax dropped from 29.3% in 2012 to 23.5% in 2015, a drop of 6%, showing that the public increasingly support an energy and environmental tax system.

Between 2012 and 2015 this survey shows that there was a relatively united and significant change in people's attitudes in Taiwan on a number of issues including low carbon energy choices, 'willingness to pay', and green tax. With the international climate change treaty bringing vertical pressure externally and the local climate related destruction also clearly evident, the public have been led to possess a concrete and significant perception of climate change. Domestically, due to people in recent years facing choices about the form of energy they want to see their country pursue, subsidies given to support the petrochemical industry and the seriousness of the air pollution situation have meant that the majority of people increasingly have the desire to choose a new development path. These risk perceptions reveal a change in the attitude of society as a whole, both in terms of preferences and choices, to a certain extent representing a new perception and value trend and important changes to cultural components (Beck 2015). Such a change in structure has become an important place of reference for explaining social structures and changes.

The triple helix movement shows that overall society's perception and value is at the cutting edge of structural transformation stemming from knowledge of the impact of energy CO_2 levels, energy consumption and pollution, industrial transition, and the effects of air pollution on health.

However, the despite the gradual unfolding process of a triple helix energy transition, while society is experiencing reform and attempts to support the development of renewable energy have increased, the long-term dominance of the nuclear complex still exists. Even though the KMT regime passed a renewable energy development bill in June of 2009, Taiwan's renewable energy is still very low as a percentage of all energy produced. Globally most industrialized nations' renewable energy accounts for an average of 13% of all generated energy, while between 2001 and 2015 excluding conventional hydro, Taiwan's renewable energy accounted for only 2.33% of all generated power, having increased by 1% in the past 15 years (Energy Bureau 2015). And besides the nuclear complex and the many structural limitations that it poses, there is also the fact that a pro-nuclear camp has advocated nuclear power as being the best option for replacing high CO_2, high energy consumptive powers. However, this explanation has been criticized by many energy democracy academics (Chou 2016b; Liu 2016; Zheng 2016), even pointing to the potential of the development of successful examples of renewable energy spread out around Taiwan (S. F. Gao 2016; Lin 2016; Fang 2016; Y. F. Chen 2016); while the 'Fun Green Electricity Alliance', 'Taiwan Renewable Energy Alliance', 'One Person a Thousand Watts' and many other civil power plants and green spots, and even Taiwan's environmental protection alliance have been promoting the development of forms of renewable energy throughout Taiwan.

These developments, show a new renewable energy discourse, a new force breaking through the old. In recent years these events have taken place on the stage of Taiwan's energy transition. Discourse and proposed ideas, as well as the practical promotion of practical energy generation will gradually attract the attention of Taiwan's public, whether it be the forming of local energy companies, the promotion of communities' own civilian power plants, biogas power generators, energy co-operatives, helping the old or orphanages to build solar panels, promoting temples around the countries to attach solar panels, or developing many other diverse and pluralistic methods to increase local capacity. Or to put in another way, increasing Taiwan's renewable energy discourse is increasing Taiwan's energy fuel choice opportunities in the energy transition movement, with one helix connecting to the next.

On 31 May 2016 Taiwan's temperature reached 38.7° Celsius, breaking a 120-year record and leading to TaiPower's operating reserve only having 1.64% left, only narrowly avoiding a power shortage crisis; for the recently elected DPP such a major emergency occurring less than two weeks after taking power meant that their response was something of a mess. The new cabinet expressed plans to restart Nuclear Plant 1's first nuclear reactor. This suggestion was met with deep dissatisfaction from those environmental groups who have worked with the DPP closely over a number of years, and who believe that the real issue is with the mismanagement of it power dispatcher rather than there being a problem with the installed capacity. They appeared at court the following day to make charges against the new government. From another perspective, the dispute over the power shortage crisis was an opportunity for supporters of nuclear power to criticize the DPP claiming their non-nuclear garden policy was unrealistic,[10] including the industry (L. C. Chen 2016; Lei 2016) and nuclear engineering scholars (Ye 2015; Jiang 2016; Li 2016), all of whom sent letters to the media, arguing that the plans for fourth nuclear power plant should be reassessed and the importance of the already running nuclear generating stations 1 and 3. Following this, Taiwan's renewable energy Promotion Alliance questioned whether Taipower had underestimated the solar panel generated capacity, as a result of which they cause a power shortage scare (R. P. Gao 2016). Anti-nuclear groups even criticized Taipower's long-term manipulation of the figures and began a new wave of Open Taipower information on the Internet, demanding that Taipower should change its long-term support for nuclear power and support the new government's policy to eradicate nuclear power.

On the first energy fuel choice helix movement, we can see that domestically a tension relationship exists, in particular, various groups possess different social technical imagination about different forms of energy and this spreads to the debates on support or denial of energy intensive industries. And all of this reveals the deep lack of trust and the competition between energy and industrial paradigms. Despite this, the Fukushima accident was an important point to lead and bring about a flip in social perception; and the new government's policy goals of a nuclear-free homeland and in support of the development of renewable power, structurally challenge the nuclear complex's intertwined power.

Discussion and conclusion

Due to the historical structure in which it has developed, the three helix movement of Taiwan's energy transition – energy fuel and CO_2 emissions, industrial transformation, and air pollution governance – is essentially a global trans-boundary risk movement, involving vertical pressure from International Green Treaties and horizontally from social cultural pressures. In the past 30 years these various pressures have become mutually entangled and agitating. The trans-boundary nature of these risks means that the development and influence of each helix is difficult to contain, surpassing the framework of its own original issue, influencing and interacting with the other helixes, while on a geographical level crossing boundaries, with its impact stretching to other regions and even globally. Therefore, no matter whether it be the long-term dominance of the nuclear complex, the movement against high energy consumption and CO_2 emissions, PM2.5 air pollution, or even renewable energy, all these issues possess different social technological imaginations, with competitions raging between the possibilities of high carbon or low carbon economic systems, sustainable value and so on.

In the aftermath of the Fukushima disaster we can see that a once weak and droopy energy transition process has been given new life, and with this has seen the introduction of new components. In particular, each helix possesses a robust social participation and challenge to technocratic expert politics, constructing a new democratic policy-making path. And this development, as Beck (2015) shows in the proposal of emancipatory catastrophism, namely the outrageous nature of contemporary nuclear disasters, dramatic climate change, and severe air pollution, has created the liberation of civil society from technocratic technology and politics. Pluralistic, diverse local groups have persisted in producing robust knowledge, revealing risk and besides attempts to break with the ignorance fundamental to the technology dominance discourse in the areas of nuclear energy, industry, and air pollution, they attempt to propose new socially sustainable goals and paths.

By taking a closer look, we can see that Taiwan's energy transition helix has experienced different historical stages, in the process creating a risk social meaning. From its beginnings in the late 1980s to its peak in the 1990s, the anti-nuclear movement has moved past its weakest point in the late 1990s. Even the attempts to adopt nuclear power as a tool against high energy consumption, high carbon emissions, and pollution in the fight against climate change have not prevented the anti-nuclear movement from presenting their case. In 2011 in the aftermath of the 311 Fukushima nuclear disaster it was able to combine with the anti-PM2.5 air movement to propose a new concept of renewable energy. This has formed a new high point in the anti-nuclear movement which continues till the present day. And the entangled trans-boundary nature of these risk helixes from a global perspective, are not only the cause of local political and economic conflicts but also are a part of a larger cosmopolitan governance. Currently, the characteristic of robust social participation is a result of a struggle with a once strong nuclear regime and complex, breaking free of its monopoly, and the rising, fading away, and re-emergence of the anti-nuclear movement in the face of all this, and this reflexively constructs competitive transformative capacities.

While this three-helix movement is one driven by civil society, no matter be it East Asia or global trends, all possess high social meaning, and have the potential to become a paradigm for risk governance globally. However until now, the reality is that structurally, Taiwan's energy transition process currently faces many difficulties.

In the past 20 years CO_2 emissions have continued to increase from 116% to 137%, with an average increase of more than 4.9%. In 1990, emissions were around 110 million tons, a figure which had increased to 252 million tons in 2008, accounting for 1% of global emissions. While the most recent figures show that in 2014 average CO_2 emissions per person have increased to 10.95 tons, in the per ten million population bracket, Taiwan was the eighth highest emitter of CO_2 globally (IEA 2015). In 2014 Taiwan's generated power structure analysis showed that petrochemical fuel (thermal fuel) accounted for 79% of all energy, with coal generated electricity accounting for 46.94%, LNG energy accounting for 28.97%, and oil accounting for 2.79%; on the other hand nuclear generated energy accounted for 16.3%, while renewable energy only accounted for 3.8% (Liou 2015; Energy Bureau 2015).

In terms of energy value, according to Key World Statistics Report (IEA 2015) it is pointed out that in 2013 the average per capita household use of electricity was 10,458 kilowatts, making it the 12th largest user of electricity per capita globally (after deducting differences regionally and organizationally), the highest of its neighbouring countries, with South Korea's average making it 13th (10,428 kilowatts), and Japan coming in at 21st (7,863 kilowatts). At the same time, in terms of subsidies for petrochemical fuel, according to the International Energy Association (IEA) statistics, in 2014 Taiwan's petrochemical fuel subsidies were as much as two hundred million US dollars annually (IEA 2015).

The structural aspects of Taiwan's current brown economies and brown energy have led to energy transition falling into the trap of a self-isolated whirlpool. Such energy subsidies, while sustaining industrial growth, have in the long term caused a lack of innovation and motivation in the industrial sector. On the other hand, loose environmental controls have caused enterprise to become accustomed to externalizing the environmental cost, with a deficiency in motivation to invest in R&D making them unable to complete globally in the long term. The difficulty of this form of rent seeking as Evans (2005) believes, is that modern developmental states need to carry out a comprehensive institutional turn, the social engineering of which is relatively vast. In particular, in East Asia, the long-term technocratic political dominance of the brown regime is now facing wave after wave of these new energy transition helix movements, being drawn into a whirlpool of impact, which could potentially deconstruct the whole.

Notes

1 Wang (2012) pointed out that the concept of reflexive governance proposed by Beck (2006) could provide a new perspective and method for analysing developmental states as they face global changes and challenges for which they need to carry out a thorough investigation of the social, economic, and environmental and such aspects affected the process of development The author believes that especially as developmental states' authoritative politics subsides, moving towards a more democratic form of governance, this concept could be used to observe the confrontation between government and society on the issues of technology, energy, economy, and environment.

2 Also see Nowotny and Leroy (2009).

3 In the face of technological expert monopolization of expert governance, many academics proposed the promotion of technological democracy in the form of risk communication, advocating long-term, sustainable, reflexive modernization perspective for resolving technological disputes, designing an open public participation and policymaking process (Beck 2006; Voß *et al.* 2009; Delvenne 2010; Marris *et al.* 2008; Rip 2005; Grin 2005).
4 This alliance calculated that in the past decade the petrochemical industry's contribution to the national GDP has remained at a level of 1.67%, a contribution which does not seem worth the relatively high proportion of energy and resources consumed in the process. cf. Chou *et al.* 2010.
5 The academic resistance group to Kuo Kuang Petrochemical Corp. (the Eighth Naphtha Cracker) founded in Changhua (2010) Press conference 'retort to statements of Ministry of Economic Affairs. Truth (Shi Hua, petrochemical) from academia'. Available from: http://www.coolloud.org.tw/node/55205.
6 In an analysis of this survey showed that the people's attitude to the harms caused by Kuo-Kuang was polarized. *China Times Daily* 2011/4/10.
7 The medical alliance stressed that Tsuang's research discovered that after the petrochemical operation, 'the PM2.5 pollution will increase the annual death rate by 339 to 565 people, as the petrochemical industry will expose people to a harmful risk environment with high pathogenic rate and high cancer rate, which is not what we will like to see; we hope our offspring can live in a healthy environment' http://www.taiwan921.lib.ntu.edu.tw/KKPT/KKE01.html. Retrieved: 2010.10.10.
8 https://www.youtube.com/watch?v=LWBVGaJuXsU. Retrieved: 2011.03.30.
9 See Ho's (2006) analysis of the DPP after their time in power and the weakening of environmental groups power. At the same time, after the failure to prevent the end to the plans for the fourth nuclear power plant in 2001, environmental groups and the DPP split on this issue, causing the movement to stagnate.
10 DPP energy policy advocates that nuclear power which currently accounts for 16% of all Taiwan's generated power should begin decommissioning in 2018, with an aim to become a non-nuclear country by 2025, And in order to make up for the loss in power which this decision will cause, renewable energy should be able to be developed immediately, with an aim to see renewable energy accounting for 20% of all Nationally generated power (New Frontiers Foundation 2014).

References

Academia Sinica Environment and Energy Research Working Group (2008). *Energy policy in response to global warming*. Taipei: Academia Sinica.

Air Clean Taiwan (2015) March on anti-air pollution, https://www.facebook.com/healthairaction/ retrieved: 26 December 2015.

Aldrich, D. P. (2013). Rethinking civil society–state relations in Japan after the Fukushima accident. *Polity*, 45(2), 249–264.

Beck, U. (2006). Reflexive governance: Politics in the global risk society. In J.-P. Voß, D. Bauknecht and R. Kemp (Eds.), *Reflexive governance for sustainable development* (pp. 3–28), Cheltenham: Edward Elgar.

Beck, U. (2010). Climate for change, or how to create a green modernity? *Theory, Culture & Society*, 27(2–3), 254–266.

Beck, U. (2015). Emancipatory catastrophism: What does it mean to climate change and risk society? *Current Sociology*, 63(1), 75–88.

Chan, C. C. (2012). Examining the impacts of petrochemical industry on human health from the epidemic perspective. *Taiwan Watch*, 12(3), Retrieved from: http://www.taiwanwatch.org.tw/magazine/pdf/v12n3-2.pdf

Chang, K. S. (2010). The second modern condition? Compressed modernity as internalized reflexive cosmopolitization. *The British Journal of Sociology*, 61(3), 444–464.

Chen, L. C. (2016, June 20). An Opinion Letter: Resolving a problem or creating a problem? Debating the shutting down of 2nd Nuclear Power Plants [Online forum comment]. Retrieved from: http://www.storm.mg/article/131037.

Chen, Y. F. (2016). Local engagement and renewable energy transition in Taiwan: the comparison cases of Yunlin and Pintung Counties. Paper presented in The International Conference on Comparative Administration and Smart Governance: Asian Perspectives, Taipei: Chinese Culture University. June 2016.

China Post (2011). The rise and fall of Kuo-Kuang Petrochemical: the nations opinions are polarized. *China Post*.

Chou, K. T. (2000). Bio-industry and social risk – delayed high-tech risk society. *Taiwan: A Radical Quarterly in Social Studies*, 39, 239–283.

Chou, K. T. (2002). The theoretical and practical gap of glocalizational risk delayed high-tech risk society. *Taiwan: A Radical Quarterly in Social Studies*, 45, 69–122.

Chou, K. T. (2007). Conflicts of technological policy and governance paradigm in a knowledge-based economy: A case analysis of the construction of the Taiwan biobank. *Issues & Studies*, 43(3), 97–130.

Chou, K. T. (2007). Reflexive discussion on new risk governance paradigm. *SOCIETAS: A Journal for Philosophical Study of Public Affairs*, 22, 179–233.

Chou, K. T. (2009). Reflexive risk governance in newly industrialized countries. *Development and Society*, 43(1), 57–90.

Chou, K. T. (2013a). Governance innovation of developmental state under the globalized risk challenges – burgeoning civil knowledge on risk policy supervision in Taiwan. *A Journal for Philosophical Study of Public Affairs*, 44, 65–148.

Chou, K. T. (2013b). The public perception of climate change in Taiwan in paradigm shift. *Energy Policy*, 61, 1252–1260.

Chou, K. T. (2015a). Cosmopolitan approach of trans-boundary risk governance in East Asia, World Congress of Risk Analysis, Singapore.

Chou, K. T. (2015b). From anti-pollution to climate change risk movement: Reshaping civic epistemology. *Journal of Sustainability*, 7, 14574–14596.

Chou, K. T. (2015c). *New challenges to climate governance – The climate civilian paradigm shift*. [Risk Society and Policy Research Center, National Taiwan University (RSPRC)].

Chou, K. T. (2016a). Climate change drives taiwan's energy transition challenge. In K. T. Chou and Z. L. Lin (Eds.), *Fourteen Lectures of Energy Transition in Taiwan* (pp. 3–26). Taipei: Chuliu Publisher.

Chou, K. T. (2016b). Beyond high carbon society. *AIMS Energy*, 4(2), 313–330.

Chou, K. T., et al., (2010). The academic resistance group to Kuo-Kuang Petrochemical Corp. (the Eighth Naphtha Cracker) founded in Changhua. In: S. Wu, M. Y. Wu (Eds.), *Wet land, petrochemical industry, island imagination* (pp. 210–213). Taipei: Route Culture.

Chou, K. T., and Liou, H. M. (2012). Analysis on energy intensive industries under Taiwan's climate change policy. *Renewable and Sustainable Energy Reviews*, 16, 2631–2642.

Chou, K. T., and Lin, Z. L. (eds.) (2016) *Fourteen Lectures of Energy Transition in Taiwan* (pp. 3–26). Taipei: Chuliu Publisher.

Delvenne, P. (2010). Parliamentary technology assessment institutions as indications of reflexive modernization. Society for Social Studies of Science Annual Meeting with JSSTS. University of Tokyo. Tokyo.

Energy Bureau (2015). *Republic of China 104 Energy Statistics Annual Report.* Taipei: Bureau of Energy, Ministry of Economic Affairs.

European Environment Agency. (2001). Twelve late lessons,.In *Late lessons from early warnings: The precautionary principle, 1896–2000.* Copenhagen: European Environment Agency.

Evans, P. (2005). The challenges of the "institutional turn": New interdisciplinary opportunities in development theory. In V. Nee and R. Swedberg (Eds.), *The economic sociology of capitalism* (pp. 90–116). Princeton, NJ: Princeton University Press.

EPA (2012). Revision of Air Quality Standard, http://ivy5.epa.gov.tw/epalaw/search/LordiDispFull.aspx?ltype=04&lname=0060, retrieved: 15 May 2015.

EPA of US (2012). Table of Historical Particulate Matter (PM) National Ambient Air Quality Standards (NAAQS) (1971–2012). Retrieved on 24 August 2016 from: https://www.epa.gov/pm-pollution/table-historical-particulate-matter-pm-national-ambient-air-quality-standards-naaqs.

Executive Yuan (2010). *Sustainable Energy Policy Action Plan.* Taipei: Executive Yuan.

Fang, S. H. (2016). The energy cooperation in the energy transition process. In K. T. Chou and Z. L. Lin (Eds.), *Fourteen Lectures of Energy Transition in Taiwan* (pp. 199–213). Taipei: Chuliu Publisher.

Funabashi, H. (2012). Why the Fukushima nuclear disaster is a man-made calamity. *International Journal of Japanese Sociology*, 21(1), 65–76.

Gao, R. P. (2016). Activity review: self-regulated climate change saloon activity. [Risk Society and Policy Research Center, National Taiwan University (RSPRC)]. Retrieved from http://rsprc.ntu.edu.tw/zh-TW/m06/ccadaptation/388-ndc-salon-0615-review.

Gao, S. F. (2016). The local practice of energy transition: Social energy and civilian plant. In K. T. Chou and Z. L. Lin, (Eds.), *Fourteen Lectures of Energy Transition in Taiwan* (pp. 181–197). Taipei: Chuliu Publisher.

Grin, J. (2005). Reflexive modernization as a governance issue or designing and shaping re-structuration. In J. P. Voβ, B. Dierk and K. René (Eds.), *Reflexive governance for sustainable development* (pp. 54–84). Northampton, MA: Edward Elgar.

Hendriks, C. M., and John, G. (2007). Contextualizing reflexive governance: The politics of Dutch transitions to sustainability. *Journal of Environmental Policy & Planning*, 9(3), 333–350.

Ho, M. S. (2006). *Green democracy: Study on Taiwan's environmental movement.* Taipei: Socio Publishing.

Hsiao, H. H. (1987). *We Only Have One Taiwan: Anti-pollution, ecological protection and environmental movement.* Taipei: Yuan Shen.

Hsiao, H. H. (1993). *Taiwan 2000.* Taipei: Commonwealth Magazine.

Hsu, H. T., et al., (2003). The health risk assessment on harmful air pollutants in Petrochemical Industrial Park. The 16th Environmental Planning and Management Meeting. Taichung: National Chung Hsin University.

International Energy Agency. (2015). Fossil fuel subsidy database, OECD/IEA. Retrieved from: http://www.worldenergyoutlook.org/resources/energysubsidies/fossilfuelsubsidydatabase/.

Jasanoff, S. (2005). *Design on nature: Science and democracy in Europe and the United States.* Princeton, NJ: Princeton University Press.

Jasanoff, S. (2010). A new climate for society. *Theory Culture Society*, 27, 233.

Jasanoff, S. (2003). Technologies of humility: Citizen participation in governing science. *Minerva* , 41(3), 223–244.

Jiang, R. T. (2016, June 13). An opinion: Do you have to be anti-nuclear to be for environmental protection? *Feng Chuan Mei*. Retrieved from: http://www.storm.mg/article/128023.

Kingston, J. (2013). Nuclear power politics in Japan, 2011–2013. *Asian Perspective*, 37, 501–521.

Lei, X. (2016). Using common sense to discuss Taiwan's nuclear problem. *Hao Doo Book Shelf*. Retrieved from: http://www.haodoo.net/?M=lui&P=11.

Li, M. (2016, June 20) Bullying our nuclear power experts, how much more of our national resources will be wasted? *Apple Daily Forum*. Retrieved from: http://www.appledaily.com.tw/realtimenews/article/new/20160620/890089/.

Lin, J. Y. (1989). The protest against nuclear power is a flight against dictatorship. Taipei: *Independent Evening Post*.

Lin, S. X. (2011). PM2.5 and your health, *Liberal Times*, 24 April 2011, Opinion Squares.

Lin, Z. (2016). Energy transition local strategy: City energy company. In K. T. Chou, and Z. L. Lin (Eds.), *Fourteen Lectures of Energy Transition in Taiwan* (pp. 165–179). Taipei: Chuliu Publisher.

Liou, H. M. (2015). Carbon emissions reduction in Taiwan's electric power industry. *Journal of Advances in Clean Energy*, 2(1), 18–34.

Liu, C. H. (2012). How to promote Taiwan's sustainable development. The 9th National Science and Technology Conference. National Science Council NSC, Taipei.

Liu, L. R., and Tang, J. L. (2011, March 17). Denying that the Environmental Groups are surrounded: the bedrock will be fragmented. *Liberty Times*, 17 March 2011.

Liu, S. B. (2016). German green energy economic transition. In K. T. Chou and Z. L. Lin (Eds.), *Fourteen Lectures of Energy Transition in Taiwan* (pp. 65–78). Taipei: Chuliu Publisher.

Marris, C. J., Pierre, B., and Rip, A. (2008). Interactive technology assessment in the real world: Dual dynamics in an iTA exercise on genetically modified vines. *Science, Technology and Human Values*, 3, 77–100.

Matsumoto, M. (2013). 'Structural disaster' long before Fukushima: A hidden accident. *Development and Society*, 42(2), 165–190.

Miller, C. A. (2008). Civic epistemologies: Constituting knowledge and order in political communities. *Sociology Compass*, 2(6), 1896–1919.

Miller, C. A. (2015). Knowledge and democracy. The epistemics of self-governance. In S. Hilgartner et al. (Eds.), *Science and democracy. Making knowledge and making power in the biosciences and beyond* (pp.198–219). London: Routledge.

New Frontier Foundation (2014). *DPP New Energy Policy*. Taipei.

Nowotny, H. (2003). Democratising expertise and socially robust knowledge. *Science and Public Policy*, 30 (3), 151–156.

Nowotny, H., and Leroy, P. (2009). Helga Nowotny: An itinerary between sociology of knowledge and public debate. *Natures Sciences Sociétés*, 17, 57–64.

Nowotny, H., Scott, P., and Gibbons, M. (2001). *The co-evolution of society and science. Re-thinking science: Knowledge and the public in an age of uncertainty* (pp. 30–49). Cambridge: Polity Press.

Rip, A. (2005). Technology assessment as part of the co-evolution of nanotechnology and society: the thrust of the TA program in NanoNed. Conference on Nanotechnology in Science, Economy and Society, Marburg, Germany, 13–15 January 2005.

Rip, A. (2010a). In a third wave of science studies? *Social Studies of Science*, 3(33), 419–434.

Rip, A. (2010b). Social Robustness and the mode 2 diagnosis. *Science, Technology & Innovation Studies*, 1(6), 71–74.

Shi, X. M. (2002). Statement by the National University Female Student Action Alliance Against Nuclear. *Taiwan Environment*, 35.

Stirling, A. (2007). Risk, precaution and science: Towards a more constructive policy debate. Talking point on the precautionary principle. *EMBO reports*, 8(4), 309–315.

Stirling, A. (2014). Transforming power: Social science and the politics of energy choices. *Energy Research & Social Science*, 1, 83–95.

Sugiman, T. (2014). Lessons learned from the 2011 debacle of the Fukushima nuclear power plant. *Public Understanding of Science*, 23(3), 267–524.

Tsai, W. C. (2010, October 7). Petrochemical pollution vulnerable to cancer medical community voices its anti-Kuo-kuang sentiment, *Liberty Times.*

Tsai, W. H. (1997). The political-economic analysis about the development of Taiwan's petrochemical industry. *Soochow Journal of Political Science*, 8, 157–224.

Tsuang, B. J., and Kuo, P. H. (2011). The air pollution and health risk of petrochemical industry. *Taiwan Watch*, 12(3).

Tu, W. L. (2016). *The undetectable toxic scent: The challenge of Taiwan's air pollution governance.* Working paper, RSPRC, National Taiwan University.

Urry, J. (2011). *Climate change and society.* Cambridge: Polity Press.

Voβ, J. P., Smith, A., and Grin, J. (2009). Designing long-term policy: Rethinking transition management. *Policy Sciences*, 42, 275–302.

Wang, J. H. (2012). Long live development studies. *City and Planning*, 39(2), 1–18.

WBGU. (2011). *World in transition: A social contract for sustainability. Flagship report.* Retrieved From http://www.wbgu.de/fileadmin/templates/dateien/veroeffentlichungen/hauptgutachten/jg2011/wbgu_jg2011_en.pdf.

Weingart. P. (2008). How robust is 'socially robust knowledge'? In M. Carrier, D. Howard, and J. Kourany (Eds.), *The challenge of the social and the pressure of practice: Science and values revisited* (pp. 131–145). Pittsburgh: University of Pittsburgh Press.

Weiss, L. (1998). *The myth of the powerless state: Governing the economy in the global era.* Cambridge: Polity Press.

Wong, C. M. L. (2012). The developmental state in ecological modernization and the politics of environmental framings: The case of Singapore and implications for East Asia. *Nature and Culture, 7*(1), 95–119. doi:10.3167/nc.2012.070106.

Yang, S. M. (2011, March 16). Nuclear power plant is a rock, Taiwan nuclear power fear of faux pas. *The Epoch Times.*

Ye, Z. G. (2015, March 10). Where will the next generation of electricity come from? *Economic News.* Retrieved from http://www.economic-news.tw/2015/03/electricity-tomorrows.html.

Ye, Z. G. (2016, June 7). Li Shi-Guang the test of a lack of power is just the beginning. *China Times.* Retrieved from http://www.chinatimes.com/newspapers/20160607000488-260109.

Zheng, J. W. (2016). The US response to climate change: Energy and economic transition. In K. T. Chou and Z. L. Lin (Eds.), *Fourteen Lectures of Energy Transition in Taiwan* (pp. 79–92). Taipei: Chuliu Publisher.

Part II

Participation and community power

4 Citizen participation-based energy transition experiments in a megacity

The case of the One Less Nuclear Power Plant in Seoul, South Korea

Sun-Jin Yun

1 Introduction

The 2011 nuclear disaster in Fukushima, Japan, has revealed the risks and vulnerability of the modern nuclear energy system. It was particularly shocking that such an enormous and tragic nuclear disaster occurred in Japan, a country that had a good reputation in terms of risk management with advanced nuclear technology.

South Korea was aggressively expanding its nuclear capacity (Yun 2015). In 2011, when the Fukushima nuclear disaster occurred, nuclear power accounted for 11.7% of the total primary energy supply and 31.3% of electric power generation in South Korea. Currently, there are 24 nuclear power reactors in operation, the oldest reactor permanently stopped on June 19, 2017, five reactors under construction, and four reactors preparing for construction but to be canceled. South Korea ranks sixth in the world in terms of installed capacity (22,529MW) and number of reactors. Due to its small land area (97,466km^2), South Korea is ranked first in the world for nuclear density, installed capacity divided by land area. Despite the Fukushima nuclear disaster, the ex-governments, Lee Myung-bak and Park Geun-hye governments, planned to build five to seven more reactors, with a total capacity of 7GW, by 2035.

Furthermore, South Korea is one of the biggest emitters of greenhouse gases (GHGs) in the world. In 2015, South Korea ranked fourth in the OECD and seventh in the world in terms of annual CO_2 emissions from fossil fuel combustion, with 586.0MtCO_2 (IEA 2017); South Korea ranked eighth among OECD member countries and fourteenth in the world in total emissions from 1990 to 2010 (IPCC 2014). In 2015, South Korea's carbon emissions per capita ranked tenth among the OECD countries and 21st in the world; its share of global CO_2 emissions was 1.81%, while its share of the global population was 0.7%. This means that South Korea emits at a rate 2.5 times higher than its population share. Its rate of increase in carbon emissions between 1990 and 2011 was the highest in the OECD. In sum, South Korea is increasingly a contributor to climate change.

South Korea ratified the United Nations Framework Convention on Climate Change (UNFCCC) in 1993 and the Kyoto Protocol in 2002. South Korea came to take a more or less proactive position during the Lee Myung-bak government, partly because of the increase in international climate change activities. Climate change has become one of the most crucial policy agendas at high-level political forums, such as the G-8 Summit, especially since the Kyoto Protocol entered into force in 2005. In 2009, right before the 15th Conference of Parties (COP15) in Copenhagen, Denmark, the South Korean government announced a voluntary target of reducing GHG emissions by 30% from the business-as-usual (BAU) levels by 2020. In 2015, the South Korean government, which at this time was the Park Geun-hye government, submitted its intended nationally determined contributions (INDC) to the UNFCCC, which targeted a 37% GHG emissions reduction from BAU levels by 2030.

However, South Korea's performance in its response to climate change has not been sufficient. According to the Germanwatch Climate Change Performance Index (CCPI), in 2016 South Korea's CCPI ranked 54th among the 58 countries whose emissions account for over 1% of global emissions. Climate Change Tracker, also evaluated Korea's INDC as being inadequate compared to its capacity and responsibility. Therefore, responding to climate change is one of the most serious challenges facing South Korea. The ex-governments planned to expand nuclear power capacity based on the argument that nuclear power is a source of wealth and convenience, as well as being crucial for continued industrialization because it provides a stable and abundant supply of high-quality electricity while emitting much less carbon in this climate change era.

In response to these serious energy-related issues, nuclear risk and climate change, energy transition experiments have been undertaken by some parts of civil society and by several local governments in South Korea. Even though most lay Koreans still believe in the necessity of nuclear power and have not withdrawn their support even after Fukushima, there is a clear trend. 'Metamorphosis' (Beck 2015) has occurred among citizens and local governors based on their awareness of nuclear risk, enhanced by the Fukushima nuclear disaster and subsequent nuclear scandals, as well as by anxiety about climate change. Some citizens and local governors have come to support and embody the concept of energy citizenship, which advocates the pursuit of safer, more ethical, and more responsible energy use. Following this concept, several social experiments have been carried out locally, the most representative example being Seoul's One Less Nuclear Power Plant (OLNPP). Seoul, one of the world's megacities, has been working since 26 April 2012, towards building an energy self-sufficient and climate-protective city through conservation of energy, improvement of energy efficiency, and expansion of renewable energy use in face of climate change and energy crises. This is a grand experiment that had never before been tried in South Korea.

This study examines how citizens in Seoul were involved in the OLNPP, an initiative that has shown impressive performance. How has the involvement of citizens been possible in a megacity like Seoul? Can the involvement of citizens be successfully maintained? What barriers lie ahead? How can we deal with these barriers? This study explores documents and statistics relevant to the OLNPP. In addition, it partially adopts the methodology of participant observation. The author, as a member of the executive committee of the OLNPP, has participated in various events and has had opportunities to observe what actually occurred.

Section 2 will explore the theoretical background and relevant concepts to explain energy transition experiments. Section 3 will take a look at the OLNPP, focusing on its background and advent. Section 4 will examine the factors that have led to the involvement of citizens, resulting in the successful performance of the OLNPP. Section 5 concludes the study, while indicating the barriers and challenges that lie ahead.

2 Theoretical background

1 Emancipatory catastrophism and metamorphosis

Ulrich Beck (1992) defined the modern society as a risk society. Modernization was expected to reduce the scale of risks humans face through the development of science and technology, as humans have increased material growth through controlling and manipulating natural resources. However, material abundance has come with increased risks, which are now embedded in modernization. Accidents in modern society have become common because they are inevitable in the processes essential to modern society.

Nevertheless, Beck argues that it is not necessary to revolutionize modern society. Rather, the risks we are facing enhance reflexivity, so we can try to pursue reflexive modernization. The first stage of simple modernization was complete absorption of the expansion of wealth and goods. The risks that inevitably accompany modernization can now lead to the second modernization, the so-called reflexive modernization.

In developing this idea, Beck (2015) proposed new concepts, emancipatory catastrophism and metamorphosis. According to Beck, citizens who witness and experience catastrophe resulting from risks manufactured by the process of modernization are awakened, and in a sense are reborn. That is to say, citizens pursue emancipation by attempting to escape from the social conditions that cause catastrophic disasters. Through this process, the individual fundamentally changes the way he or she thinks and acts.

In particular, if there is a recognition or reflection that such a catastrophe resulted from 'organized irresponsibility', defined as a situation in which decision-makers do not take any responsibility for the affected and the affected have no specific way to engage in decision-making processes (Beck 2015), citizens come to pursue social transformation in order to enhance the safety and happiness of their family and themselves. There is a crisis of responsibility involved in modern risks because they are manufactured, or made by humans. Organized irresponsibility is connected to responsibility crises because people who are not directly involved with the cause of the risk have to take responsibility for its effects. Because of increasing uncertainty and complex causal relations, it is very difficult to figure out the source of responsibility. This situation can be identified as organized irresponsibility.

However, Beck argues that good results can come from unintended bad results. By experiencing disasters, citizens recognize problems and attempt to solve them with self-reflection. This is emancipatory catastrophism. People who pursue the process of emancipatory catastrophism and social change as a group can experience social catharsis. Social catharsis is not an automatic process following a catastrophe. Social catharsis is achieved when people sympathize with the pain of others resulting from a catastrophe, heal the wounds, and pursue social transformation beyond organized irresponsibility. Individuals who have experienced emancipatory catastrophism and social catharsis think and act differently. Beck called this change metamorphosis.

The concepts of emancipatory catastrophism and metamorphosis are useful in explaining the expansion of post-nuclear movements and energy transition movements in South Korea after the Fukushima nuclear disaster. Many people came to realize the risks associated with nuclear power, or became concerned about the sources of the energy they consumed. Their indifference to the problems associated with the energy they used disappeared. Witnessing pain caused by Fukushima made Koreans recognize nuclear risks as their own problems, because their families could be affected by these risks. They came to realize the necessity of change and started engaging in the process of change.

2 Energy citizenship

The Fukushima nuclear disaster made people start to recognize serious risks involved in nuclear safety and to promote changes in energy use. The Chernobyl nuclear accident in 1986 was as not well known in Korea because the press and civil society capacity were behind the times. Even those Koreans who did know of it did not fully recognize the risks associated with nuclear power. Fukushima, however, happened to the neighbour next door. Japan was regarded as the most advanced country in the world in terms of nuclear technology and risk management. The incident caused concerns over nuclear safety to become much more widespread in Korea. People came to recognize that nuclear risks could be closely connected with their everyday life.

In light of this, it is important to understand the concept of energy citizenship. Energy citizenship is defined as citizens' rights and responsibilities concerning energy use (Devine-Wright 2007). In other words, it recognizes that citizens have a right to enjoy the essential energy services required for a meaningful quality of life, but also must take responsibility for the social and environmental consequences created by the production, distribution, and consumption of energy. Citizens exercise the right to enjoy energy services, but in return they should recognize the negative impacts of energy production and consumption on society and nature and fulfil their responsibility to reduce them.

Citizens can transform themselves into producers by practising energy conservation, energy efficiency improvement, and renewable energy use. They can engage in energy saving movements, sometimes called negawatt movements, under the slogan 'saving is production.' They can install solar PVs on their rooftops and windows and invest money to create citizens' energy cooperatives. They can become energy farmers and prosumers, meaning simultaneous producers and consumers. These are examples of the kind of metamorphosis that Beck conceptualized.

In South Korea, the number of energy citizens has increased since the Fukushima nuclear disaster. The Fukushima accident, working as an event of emancipatory catastrophe, provided a chance for citizens to undergo metamorphosis; thus, increasing numbers of citizens have recognized the risks of nuclear technology and become involved in energy transition movements. What changes in Korean society caused this and what changes have occurred as result? The following section will explore these questions.

3 Understanding the One Less Nuclear Power Plant in Seoul

1 Key Energy Statistics for Seoul

Seoul's primary energy supply accounted for 4.2% of the national total in 2011, just before the OLNPP was initiated. In terms of the sources of primary energy supply, petroleum accounted for the biggest share (52.4%), followed by Liquified Natural Gas (LNG) (44.0%), as shown in Figure 4.1. This is somewhat different from the energy source ratio nationwide. At the national level, petroleum accounted for the biggest share (38.2%), followed by coal (30.0%), LNG (16.8%), and nuclear (11.7%).

In terms of final energy consumption by source, petroleum accounted for the biggest share (38.9%) in Seoul, followed by LNG and town gas (29.7%), and electricity (26.0%), as shown in Figure 4.2. At the national level, the share of petroleum (49.5%) was much higher than that of Seoul, while the share of

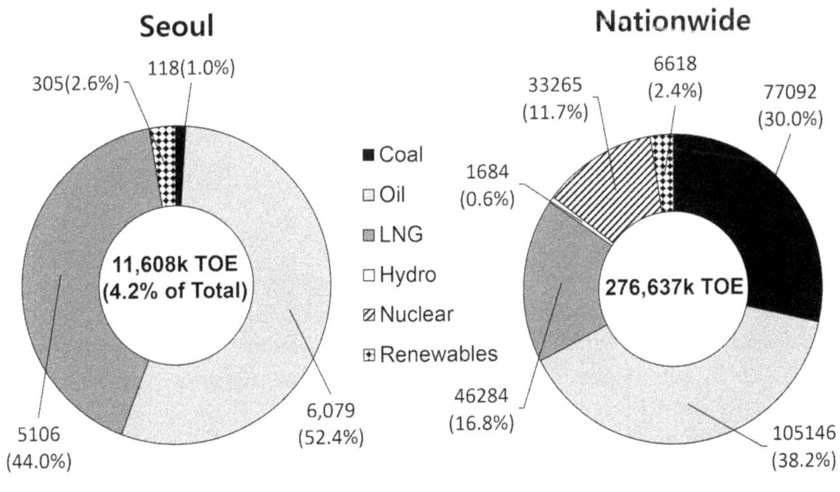

Figure 4.1 1: Primary energy supply in 2011: Seoul vs. Nationwide

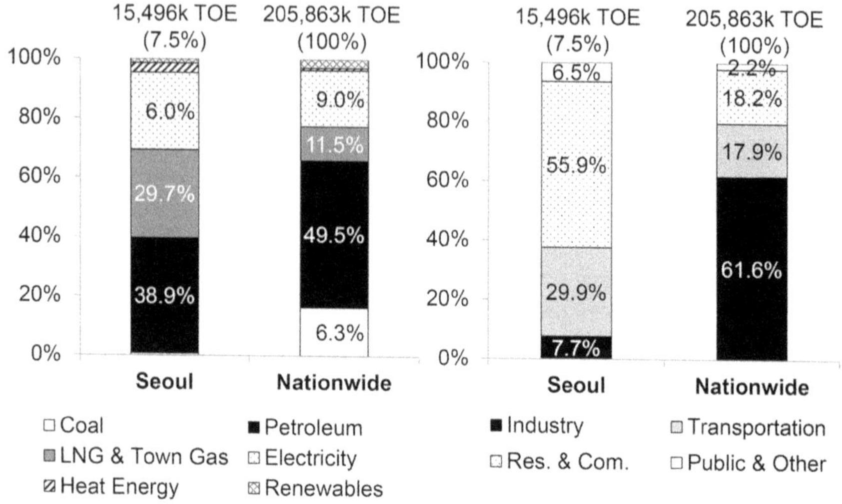

Figure 4.2 Final energy consumption by source and sector in 2011: Seoul vs. Nationwide

electricity (19.0%) was somewhat lower than Seoul's. This difference arose from the composition of the sectors. The residential and commercial sector comprised the biggest portion (55.9%) in Seoul, while the industrial sector comprised the biggest portion (61.6%) at the national level. In Seoul, the consumption of LNG and town gas and electricity was relatively high, while petroleum comprised almost half of the final energy consumption at the national level. In terms of sector, consumption by the residential and commercial sectors and transportation sector was relatively high in Seoul, while the industrial sector was overwhelmingly dominant nationwide. This implies that more effort has to be given to the residential and commercial sector and that LNG and town gas and electricity have to be prioritized for energy consumption reduction in Seoul. In the case of electricity, the residential and commercial sector consumed 82.9% in Seoul, while the industrial sector consumed 53.2% at the national level. Buildings and transportation consumed 90.9% of energy in Seoul in 2011.

2 The background of the advent of the OLNPP

What was the social context of the advent of the OLNPP? There were at least four crucial events. The first event was the Fukushima nuclear disaster in Japan in 2011. Through this disaster, the Korean public has come to recognize the risks associated with the use of nuclear power and the importance of safety in the use of energy. Public support for nuclear power in Korea was relatively high before the Fukushima nuclear disaster, but it has since decreased.

Public opinion polls conducted by the Korea Nuclear Energy Promotion Agency (KNEPA) show changes in public support before and after the Fukushima nuclear disaster. KNEPA conducted opinion polls bimonthly before the

Fukushima nuclear disaster, but it did not conduct opinion polls in 2011 until October. In the last opinion poll before the Fukushima nuclear disaster, conducted in November 2010, 89.4% of the respondents agreed with the necessity of nuclear power. In the same poll, 45.9% supported an increase in the number of nuclear power plants, 43.0% agreed with maintaining the current number, and only 9.3% supported decreasing the number of nuclear power plants. Concerning the construction of nuclear power plants within their community, 27.5% agreed, 60.2% disagreed, and 10.2% said it would depend on the scale of investment in the community. With regard to the safety of nuclear power plants, 53.3% answered that they were safe, while 41.4% answered that they were risky.

However, the first opinion poll after the Fukushima nuclear disaster, conducted in October 2011, showed different results. Now only 78.3% of the respondents agreed with the necessity of nuclear power. Concerning the number of nuclear power plants, 37.4% supported an increase, 38.4% agreed with maintaining the current number, and 15.7% supported a decrease in their number. 0.4% of the respondents preferred the closure of all nuclear power plants. Concerning the construction of nuclear power plants within their community, 20.9% agreed, while 78.4% disagreed. With regard to the safety of nuclear power plants, 39.5% answered that they were safe, while 57.9% answered that they were risky.

Distrust in the safety of nuclear power plants was exacerbated in December 2011, when procurement corruption and bribery were revealed to the general public. Secondhand equipment for nuclear power plants was delivered as if new. High-level as well as entry-level workers at the Korea Hydro and Nuclear Power Company were arrested on charges of bribery.[1]

The second event was the increasing threat driven by climate change. Seoul has experienced temperature increases and a rapid increase of summer precipitation. Thus, it has frequently suffered from extreme weather events including heavy rainfall and consequent landslides and inundation of roads, buildings, and cars. Citizens and policy-makers in Seoul came to recognize the seriousness of climate risk, not from an abstract dimension but from an empirical one. Climate change has become a visible disaster. The images of flooded roads, including Sejong-ro, the symbolic central road in Seoul, Kangnam-ro, one of the most popular roads, and Daechi-ro, one of the richest communities' roads, strongly impacted citizens' minds. Climate change is not an issue distant from citizens' daily lives. The summer of 2011 saw the first victims of landslides resulting from heavy rainfall in Seoul since independence from Japanese colonialism. Several deaths even happened in the rich communities. Recently, repeated extreme cold waves in the winter and extreme heat waves in the summer have also reminded citizens of the seriousness of climate change. Citizens came to recognize that climate protection required fundamental changes in the production and consumption of energy.

A circular blackout on 15 September 2011 was the third event. Blackouts are rare in Korea, and for the first time the blackouts occurred due to a shortage of power supply, rather than a natural disaster, accident, or mechanical trouble (S. Lee and Yun 2013). On that day, the amount of electricity being consumed significantly increased because of the late summer heat, while 25 power plants had been shut down for inspection and maintenance. When the reserves fell to as low as 240,000 kW, the Korean government abruptly put a circular blackout in effect to prevent a nationwide power failure. As a result, citizens in Seoul and other areas experienced the power cut without any advance notice or preparation. Almost all services, including administrative, educational, medical, financial, commercial, transportation, and communication functions, were paralysed. This reminded the citizens of the vulnerability of cities, which depend on electric power supplied from outside of the city. Energy self-sufficiency came to be recognized as a pivotal task for a sustainable life in cities.

Furthermore, the aforementioned climate change places more burdens on energy security. On the one hand, more extreme weather events like cold and heatwaves cause increased energy demand for heating and cooling, which leads to a supply shortage. On the other hand, increasing extreme weather events can also provoke unstable supply as power generation systems can be damaged due to heavy rain or typhoons. The abrupt blackout made citizens aware of the vulnerability of cities with low energy self-sufficiency, especially given the increasing climate threat.

Last, but not least, there were social conflicts around the construction of high-voltage transmission lines. In South Korea, the Capital area accounts for 40% of the maximum power load, but most electric power plants are located in other areas. Thus, with this structure, long-distance transmission is necessary. The Korea Electric Power Company (KEPCO) prefers high-voltage 765kV transmission lines to reduce power loss in transmission. However, these are accompanied by a controversy over electromagnetic waves. Residents of Miryang had fought against the construction of high-voltage 765kV transmission lines in their village since 2006 (Yun 2015), but this was not much reported in the mass media (H. Lee and Yun 2013). In January 2012, however, an old man, Lee Chi-woo, burned himself to death at a protest against the construction project and, consequently, the story of their fight was more widely reported. Some citizens began to think about the 'ethical' dimension of the centralized electricity system, which divides producing areas from consuming areas, and shifts economic, environmental, and health burdens to the small number of producing areas and villages that transmission lines pass through.

Seoul is one of the world's megacities, with a population of over 10 million people – its population was 10,528,774 in 2011. In 2011, Seoul consumed 46,903GWh of energy, accounting for 10.3% of the total electricity consumed nationwide (455,070GWh). However, Seoul generated only 1,384GWh, which accounted for only 2.95% of the electricity it consumed and was only 0.28% of the total electricity generated nationwide (496,893GWh). This shows that Seoul relies heavily on other regions of the country for electricity, while at the same time one-sidedly placing economic, environmental, and health burdens on these other regions. In particular, nuclear energy produced 31% of the electricity generated that same year. Increases in electricity consumption in Seoul lead to increases in the production of electricity from nuclear power. In the midst of the increasing awareness of nuclear risk and the decreasing public acceptance of nuclear power due to the Fukushima nuclear disaster in 2011, Seoul launched an initiative to deal with the risks associated with energy and climate change, named 'One Less Nuclear Power Plant.'

3 Contents of the OLNPP

Through the OLNPP, Seoul aimed to eliminate two million tons of oil equivalent (MTOE), which is comparable to the amount of electricity generated by one nuclear power plant, from its use of energy through conservation, efficiency, and the production of renewable energy by the end of 2014. Seoul also planned to improve its self-sufficiency in electricity, which is the generation of electricity divided by the consumption of electricity, from 2.95% in 2011 to 8% by 2014 and 20% by 2020.

In April 2012, the Seoul Metropolitan Government (SMG) announced its 'Comprehensive Plan for One Less Nuclear Power Plant (2012–2014).' This plan included 78 projects in three areas: new and renewable energy production, more efficient energy use, and saving energy. The city government introduced various policies and institutional arrangements to accomplish the target. As of June 2014, Seoul had already accomplished the targeted reduction of total energy consumption, 2 MTOE, six months earlier than the original timeline.

SMG anticipated achieving this target ahead of schedule and began discussions on the second phase of the initiative in January 2014. The Implementation Council for the OLNPP, a public-private governance organization, led discussions about the values and vision for Phase 2. It made draft proposals in which the implementation tasks for Phase 2 were identified, followed by a forum on energy policies to collect the opinions of experts and citizens. Finally, SMG initiated the second phase of the OLNPP in August 2014.

The first phase of the OLNPP focused mostly on quantitative target management. In the second phase, more focus was given to the values embedded in energy, including energy self-sufficiency, energy sharing, and energy participation. Improvement of self-sufficiency in energy means decreasing dependence on outside energy sources and transitioning into a city responsible for its own energy consumption, which leads to a city secure and safe from energy crises. Furthermore, it enables more opportunities to grow energy industries and to create jobs. Energy sharing means sharing energy with more vulnerable groups and future generations. It is expected to lead to an equitable distribution of the production and consumption of energy and also the expansion of energy welfare to the energy poor. Finally, energy participation means the establishment of energy governance that is open to citizens for energy policy making and practice. In addition, energy participation includes disclosing energy information and policies and providing opportunities for education and training.

Targets of the second phase of the OLNPP are saving and producing energy to substitute two nuclear reactors, equivalent to 4 MTOE; improving Seoul's electrical self-sufficiency to 20%; and reducing 10 million tCO_{2eq} emissions by 2020, a 20.5% reduction compared to 2011, when Seoul's GHG emissions were 49 million tCO_{2eq}. The 2020 goal is 38.96 million tCO_{2eq}. Reduction of 10 million tCO_{2eq} emissions means a reduction of 1 tCO_{2eq} emissions per person. Figure 4.3 summarizes the essence of the OLNPP 2.

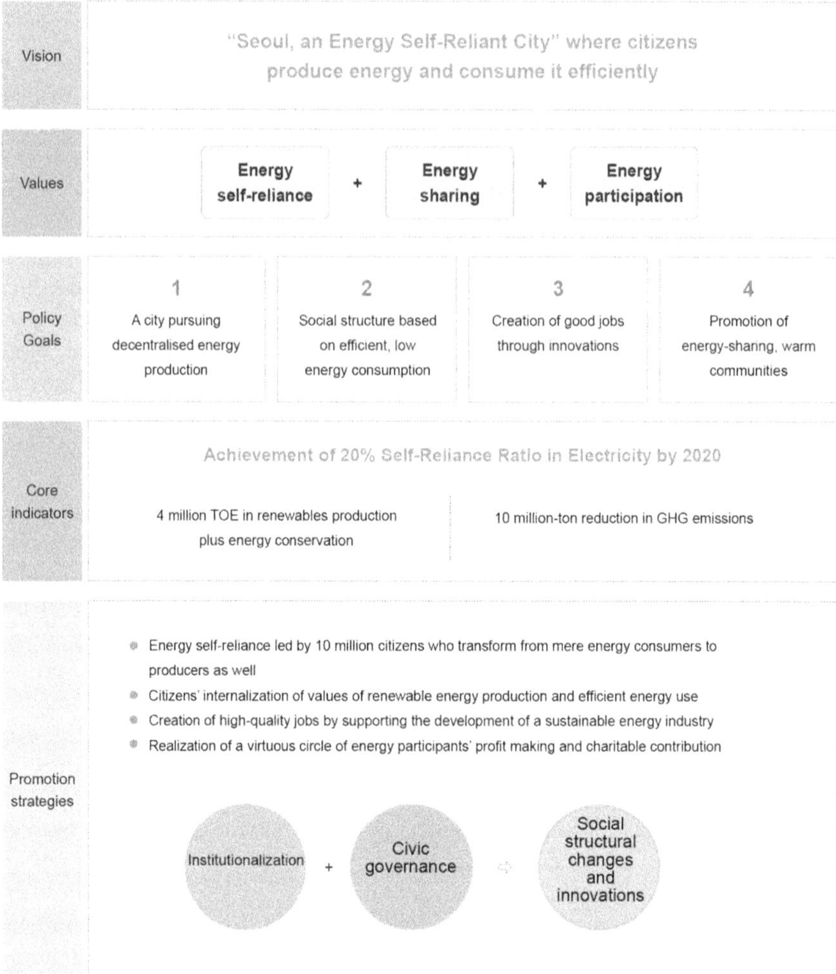

Figure 4.3 Diagram of vision, values, policy goals, core targets of the OLNPP 2
Source: Seoul Metropolitan Government 2014, 'One Less Nuclear Power Plant, Phase 2', p. 30.

The OLNPP 2 slogan is 'Seoul, an Energy Self-Reliant City where citizens produce energy and consume it efficiently,' based on three main values of self-sufficiency, sharing, and participation. To realize these policy goals, SMG prepared 88 projects under 23 tasks in four categories: production, efficient use and conservation, industrial jobs, and community welfare. SMG established a road map with specific annual targets by year as shown in Table 4.1.

In order to achieve the OLNPP 2 targets, SMG made institutional arrangements. SMG created a feed-in-tariff, under which power producers with small-scale PVs under 50kW are supported 100 with Won/kWh. Rental prices were cut down to 20,000 Won/kWh, much lower than the central government's guideline

Table 4.1 Major policy indicators of the OLNPP 2

Classification	*Indicators*	*Unit*	*Total*	*2014*	*2015*	*2016*	*2017*	*2018*	*2020*
General	Electricity self-sufficiency	%	20.0	5.0	7.0	9.0	12.0	15.0	20.0
	Renewables production	%	5.0	2.0	2.5	3.0	3.5	4.0	5.0
	Total energy production & reduction (accumulated)	1,000 TOE	4,000	200	900	1600	2300	3000	4000
	Total CO_2 reductions (accumulated)	1,000 TOE	1,000	15	100	300	470	660	1000
Production	PV power plants	MW	105	24	21	20	20	20	40
	Hydrogen fuel cell plants	MW	195	41	34	40	40	40	100
	Total non-utility building cogeneration (accumulated)	MW	61 (150)	1 (90)	10 (100)	10 (110)	20 (130)	20 (150)	54 (204)
Efficiency & Conservation	Promotion of office & residential building BRP	1000 Bldgs.	65	10.6	13.3	13.5	13.7	13.9	16
	LED distribution (public & private sectors)	1,000 Unit	28300	5000	5705	5750	5800	6000	11580
	Total vehicles in car-sharing service (accumulated)	Car	3000	1500	1800	2000	2500	3000	3500
	Eco Mileage Member (accumulated)	1,000 members	2800	2000	2200	2400	2600	2800	3000

Classification	*Indicators*	*Unit*	*Total*	*2014*	*2015*	*2016*	*2017*	*2018*	*2020*
Industry & Jobs	Establishment of co-ops & social enterprises	Unit	70	10	12	14	16	18	20
	Support for green energy tech shops	Case	210	–	30	50	60	70	80
	Creation of green clusters	Place	6	1	1	1	1	2	–
Community Welfare	Enactment of energy welfare ordinance	–	–	–	Enact	–	–	–	–
	Citizen participation in the Energy Welfare Fund	1,000 people	10	–	–	20	30	50	50
	Training of energy social workers	Person	180	10	20	50	50	50	50
	Creation of energy self-reliace	Place	200	15	20	35	60	70	70

Source: SMG 2014, ‘One Less Nuclear Power Plant, Phase 2’, p. 35, reconstructed.

based on official land prices. SMG extended financing and lowered interests on loans. In addition, SMG strengthened the mandates for LED installation, the use of new and renewable energy, and the installation of building energy management systems (BEMS) in buildings larger than 10,000m^2. The OLNPP 2 demonstrates that Seoul is moving from being an energy consuming city towards being an energy producing city. This is an epoch-making event.

4 Achievements of the OLNPP

As aforementioned, the energy reduction targets of the first phase of the OLNPP were achieved six months ahead of the original schedule. There was substantial reduction of energy consumption in Seoul during Phase 1 of the OLNPP. This reduction is even more impressive when contrasted with the increase in energy consumption nationwide. Electricity consumption in SMG decreased by 1.4% in 2013 compared with 2012, while nationwide electricity consumption increased 1.76% during that period. Seoul reduced its town gas consumption by 3.54%, in contrast with a nationwide increase of 1.43%. Petroleum consumption was reduced in Seoul by 1.7%, while it increased nationwide by 2.9%. Seoul's electricity sufficiency rate grew to 4.2% in 2013 from 2.95% in 2011.

By the end of 2015, the accomplishment rates for the OLNPP 2 in the areas of energy saving and production and GHG emissions reduction were beyond their targets, at 125% and 123%, respectively. However, electricity self-sufficiency rate was not achieved. It was at 5.5%, only 79% of the year's target, 7.0%. However, the rate is still improved from 2.95% in 2011. The total accumulated installation capacity of PVs increased from January 2013 to 105.01 MW, of which mini PVs accounted for 16.9%, or 17.74 MW, in a total of 10,929 households. In 2015 alone, more than 303,000 units of lights were replaced with LEDs in the public sector.[2] Total membership in the eco-mileage programme became 1.72 million, leading to energy savings of 376.496 TOE. The total number of energy self-sufficient villages became 35 in 2014 and 55 in 2016. As a result, the OLNPP initiative has drawn international attention and come to be recognized as an exemplary energy transition experiment in a city.[3]

4 Factors influencing successful involvement of citizens in the OLNPP experiment

SMG argues that the achievements until now have been based on citizen participation. It states, 'the One Less Nuclear Power Plant initiative is the citizens' action plan on energy issues' (SMG 2014: 19). Its catchphrase is 'Citizens are energy!' According to SMG, 47% of the city's households (1.68 million) participated as members of the eco-mileage programme in 2013. Around 30,000 students also participated in the OLNPP as Energy Guardian Angels at home and school. Through public-private partnerships, 20,000 residential and office buildings implemented Building Retrofit Projects (BRP) and 6.79 million LED lights were installed. Citizens' evaluation of the OLNPP is positive. An opinion poll conducted in March 2014, found that 71% knew about the OLNPP initiative and 59% evaluated it positively, with 13% providing a negative response. How did citizens come to engage in the OLNPP? Who has been encouraged to participate in the process of planning and implementation and how has this initiative been made and implemented?

1 Deliberative energy governance

The OLNPP uniquely included citizens' significant participation in its pursuit of deliberative energy governance from the beginning. It has conducted various experiments to reflect more voices from the lay citizens and to work with them. This deliberation-oriented energy governance system is a critical path for civic participation.

(a) Decision-making process of the OLNPP and its participants

In South Korea, energy policy was established through a very bureaucratic top-down process. The central government establishes energy plans and policies with little involvement from local governments. South Korea has shaped and followed a 'hard energy path' (Lovins 1976), in which decision-making is monopolized exclusively by the central government. The 'Framework Act on Low Carbon, Green Growth' (FALCGG) requires the central government to 'establish and implement a basic plan for energy every five years for a planning period of 20 years' (Clause (1) of Article 41).

Whenever the government intends to establish or amend a basic energy plan, it must present a proposed plan to the Energy Committee under Article 9 of the Energy Act, and then to the Presidential Committee on Green Growth and the State Council sequentially for deliberation. The Energy Committee and the Presidential Committee on Green Growth (PCGG) were established to work as deliberative governance entities, including actors from the civil sector. However, their roles and functions are more perfunctory than substantial. The Energy Committee seldom calls a meeting. As it has meetings for deliberation just before an energy plan is passed to the PCGG, deliberation is very limited. There have been no changes made by the Energy Committee.

According to Article 9 of the Energy Act, 'The Government shall establish the Energy Committee to deliberate on matters concerning major energy policies and energy-related plans.' Concerning committee members from the civil sector, the Energy Act states that 'Persons with a wealth of knowledge and experience in the area of energy who are commissioned by the Minister of Trade, Industry, and Energy shall become commissioned members of the Committee' (Clause (5) of Article 9). As a result, while five committee members were recommended by civil organizations, the current Energy Committee, has only one such member. Committee members from civil society are mainly professors, 10 out of 19 members. Five members are from the business sector and three members are from government-affiliated research institutes. Therefore, it does not reflect the voices of lay people and civil activists well enough.

According to the Energy Act, each Special Metropolitan City Mayor, *Do* (Province) Governor, or the Governor of a Special Self-Governing Province is required every five years to formulate and implement a local energy plan lasting not less than five years in order to efficiently achieve the objectives of the basic national energy plan, taking into account the regional characteristics of the area under his or her jurisdiction (Article 7). Generally speaking, however, local energy plans have not been seriously established and they have not been specific enough to consider regional characteristics. In addition, those plans were established mainly by officials in local governments with some comments from experts or through outsourcing.

Seoul's OLNPP, however, was different from the beginning. SMG has made efforts to collect and reflect citizens' ideas and tried to mobilize citizens' actions in those processes. In particular, the city of Seoul tried to make those participatory events deliberative. Before the Comprehensive Plan for the OLNPP was announced on 26 April 2012, 16 meetings were held from January to April 2012 to create a draft. Seoul officials of the Hope Policy Advisory Group, and diverse civil organizations worked together over four months. Based on the first draft, opinions and suggestions were collected through a 'Policy Listening Workshop on the OLNPP' on 21 February 2012, and a 'Citizens' Grand Seminar (a large-scale town hall meeting)' on 16 April 2012. Around 400 citizens participated in the Grand Citizens' Seminar, in which participants were divided into 22 groups for discussion. They suggested 109 ideas for actions, including concentrated education targeting at homes and schools, disclosure of energy consumption information from energy guzzling buildings, and the creation of more pedestrian- and public transportation-only streets. These ideas were reflected in the revised version of the draft OLNPP.

For the second season of the OLNPP, starting 20 August 2014, there were various events to collect citizens' and experts' ideas and opinions. These participatory and deliberative events can be divided into four categories: citizens' participation through forums, citizens' opinion collection through on-line systems, experts' opinion collection though the Seoul International Energy Advisory Council and the Domestic Expert Advisory Council, and a forum of energy industry employers.

There were two direct participation processes, the Citizens' Grand Seminar held on 12 March 2014 and the Open Forum with Citizens held on 26 April 2014. The new concept of a citizens' seminar was applied during the first seminar, the so-called Dream Party through Social Fiction. Its slogan was 'If we dream together, the dream can become reality.' Like writing a novel, citizens dreamed of their future together and discussed what should be done to make the dream become reality. Around 400 citizens from civil organizations, religious groups, educational groups, and diverse associations participated in this social fiction event. They had group discussions to suggest ideas about energy production, efficiency and conservation, economy, and community. The open forum with citizens was named the Forum on Hearing Policy from Citizens, in which 200 citizens from university students associations, Earth Hour Volunteers groups, energy auditors and designers, domestic experts, and foreign experts from the University of

Delaware, WWF, UNEP, Greenpeace, and the UK Embassy participated. In addition, some participants who had engaged in the Social Fiction Forum in March joined.

Citizens could participate in shaping the second stage of the OLNPP on-line. The city government invited citizens to join the second stage of the OLNPP for its re-naming. In total, 143 names were suggested (97 names offered by lay citizens, 46 names offered by city officials). Then citizens were asked to choose the name they preferred the most; around 2,000 citizens participated in the survey. Mobile voting (mVoting) was also conducted to collect public opinion on the name and slogan of the second stage of the OLNPP. Four naming options were suggested, and among them 'One Less Nuclear Power Plant: Seoul Good Energy Project' gained the most votes. Five options were provided for the slogan, and 'Energy revitalized by citizens makes citizens alive!' gained the most votes.

The direction of the second OLNPP has been established by the One Less Nuclear Power Plant Implementation Council. The OLNPP Implementation Council had plenary meetings twice and its coordination division met six times. There were 13 individual division workshops. In addition, domestic and international experts were invited for their advice. In 2013, the city government launched the Seoul International Energy Advisory Council (SIEAC), and invited ten globally renowned energy experts, including Amory Lovins. They had an international seminar on 13 November 2013 and announced the Seoul Declaration with its nine recommendations. Their advice was reflected in the plan for the second stage of the OLNPP. Citizens were invited during every stage of initiating both the first and second OLNPP.

(b) Ways citizens and experts participated in the OLNPP

SMG has set up an administrative structure for the OLNPP. First of all, a new office exclusively responsible for the OLNPP campaign was established. In April 2012, the city government established a department in the Climate and Environment Headquarters to promote the OLNPP campaign. In May 2012, it launched the Climate and Energy Research Center in the Seoul Institute with eight experts in the area. The Environmental Policy Division and the Green Energy Department have worked together to forge and implement OLNPP policies. The Green Energy Division, established in 2012, consists of six teams including the Energy Policy Team, the Sunlight Power Generation Team, the Community Energy Team, the New and Renewable Energy Team, the Building Retrofit Project (BRP) Team, and the Electric Power Management Team (Mayor of Seoul Metropolitan Government, 2013). In addition, the city government organized the Energy Citizen Cooperation Group to promote citizen participation in the OLNPP. The head of the Energy Citizen Cooperation Group was recruited from a civil organization. She was a former Secretary General of the 'Korea NGOs' Energy Network.'

Besides these governmental offices, the city government also launched two entities for participatory and deliberative energy governance, as shown in Figure

4.4 below. One of them is the One Less Nuclear Power Plant Citizens' Council, launched on 25 April 2012. The council is composed of 25 members, including the Mayor and the representatives of environmental and energy organizations, the enterprises, the religious communities, and the educational organizations; it is co-chaired by the Mayor of Seoul and the two other civil representatives. The other one is the One Less Nuclear Power Plant Implementation Council, initially composed of 45 members and later increased to 58, working for joint private-public cooperative governance and communication. Council members were drawn from civil organizations, enterprises, religious groups, educational organizations, journalists, and cultural and public services. The implementation council, co-chaired by the Deputy Mayor for Climate and Environment and a person from the private sector, orchestrates and coordinates all projects of the OLNPP. It gives advice on overall projects, picks possible projects, designs and evaluates projects, suggests ideas to speed up projects and supports individual projects that will be implemented through citizens' power. The implementation council was divided into four teams, – production, conservation, efficiency, and communication – in order to utilize professionalism in Phase 1 of the OLNPP. In the OLNPP 2, it is composed of five sub-committees, including production, efficiency improvement and conservation, energy education and communication with citizens, energy jobs, and energy community and welfare.

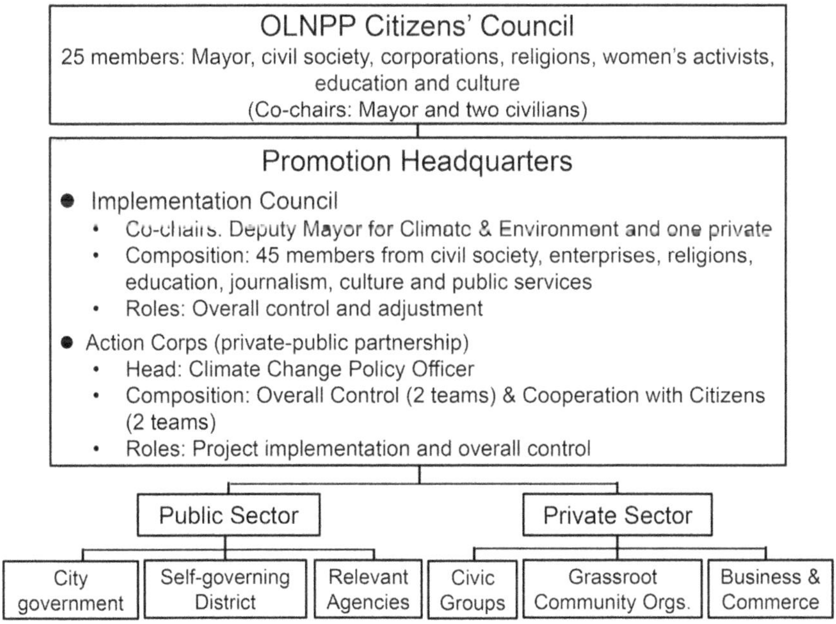

Figure 4.4 Actors and organizations involved in the OLNPP
Source: The Mayor of Seoul Metropolitan Government 2013: 12.

Implementation of the OLNPP has been designed to be based on citizen participation. The OLNPP Citizens Council aims to realize governance so that diverse participating groups are the way to make an energy production city beyond an energy saving city (the OLNPP Implementation Committee, 2014). The OLNPP has tried to create a culture of citizen-initiated energy saving movements. One of the notable citizens' action programmes is eco-mileage, which offers incentives to energy saving households. Its membership has grown from 0.5 million in December 2011 to 1.68 million in June 2014, leading to the reduction of 440 kTOE and 680 ktCO_2. Citizens can practise energy savings in their daily life based on recommendations and tips provided by the city. There is a special event held regularly. The 22nd of each month features a lights off hour starting at 8:00 pm. Families are encouraged to join the event and it can also help for them to have more time for intimate conversation.

Furthermore, there are citizens who installed PVs on their rooftops and mini PVs on their apartment balcony. To ease and support citizen-initiated small-scale PV installation, the city government improved institutional arrangements so it has been possible to offer installation sites with lowered rental fees for citizens' cooperatives, shorten the permission period from 60 to 30 days, provide long-term low-interest loans for PVs up to 150kW, and support Renewable Energy Certificate (REC) sales through a subsidy, 50 won (100 Won since 2015) per 1kWh, for five years. Citizens are encouraged to join the energy self-sufficiency village project. An energy self-sufficiency village is one that improves its energy self-sufficiency rate through reduced energy consumption based on conservation and efficiency improvements and increased energy production by expanding renewable energy. Any residents organization with more than three memberships can apply for this project. The number of self-sufficiency villages has grown from three in 2012 to 55 in 2016.

Citizens can use an energy clinic service to audit energy waste at home and workplaces. Energy auditors and consultants visit each house and workplace to provide monitoring and consulting services. These energy designers and consultants have been trained by the city government. This approach can contribute to job creation. Three cooperatives of energy designers and consultants have been established.

In schools, there are Energy Guardian Angel Corps composed of those who voluntarily save and monitor energy consumption. Currently, there are around 30,000 energy guardian angels in the 526 schools. The city government signed a memorandum of understanding (MOU) with 34 of the 42 universities located in Seoul to promote green campus movements. In those universities, students were appointed as honorary ambassadors. They volunteered to deliver information relevant to the OLNPP initiative to students, monitor university energy consumption activities, and practise diverse energy saving activities. These 34 universities have established the Seoul Green Campus among themselves to share information and experiences. Also, the city government opened an educational course called the 'Green Leadership programme' for working-group officials who are in charge of managing facilities on university campuses.

There is even voluntary participation among the business and religious sectors. Large supermarkets, construction companies, associations of hospitals, and religious groups have set voluntary energy conservation goals and started energy-saving activities. For example, there is a project called a 'good-nature store project' targeted for medium- and small-scale stores. As energy consumption by the commercial sector in Seoul has been increasing, the city government has encouraged stores to voluntarily reduce energy consumption and supported various strategies via energy designers and NGOs. If those stores accomplish their energy saving goals, the city government gives a 'good-natured store' mark and gives incentives. NGOs have played a role as a bridge connecting the city government and the private sector. Religious groups also begun to join the OLNPP. Some of them changed their lights to LEDs, adopted voluntary energy conservation targets, and installed PVs on their rooftops.

At the beginning of citizen involvement, those who had already experienced metamorphosis with energy citizenship participated in various projects of the OLNPP, but some citizens have gained the identity of energy citizens through the participation processes (Park and Yun 2016). Participation transformed citizens to being more sensitive to energy use and sustainable city life.

2 Political leadership

The impacts of Fukushima were not limited to the city of Seoul. With differences of degree, most of the cities and citizens witnessed and experienced these events together. So why was the OLNPP project initiated in Seoul? Furthermore, how has the OLNPP encouraged citizen involvement and implemented citizen participation? One of the main influential factors was the leadership and will of the mayor, among others. The One Less Nuclear Power Plant project was strongly supported by the mayor, Park Won-soon. Since the central government has aggressively expanded the installation capacity of nuclear power plants and maintained strong support for nuclear power even after the Fukushima nuclear disaster, it was delicate to name the local government's project that way. Seoul's OLNPP implies a direction opposite to the central government's energy policy and the central government pressured it to change the name. However, Mayor Park strongly supported the project name.

Mayor Park's individual experience is deeply associated with his attitude towards nuclear power. He was involved in the February 2004 referendum in Buan, which was voluntarily undertaken to reflect local residents' opinion on the location of a disposal site for radioactive wastes. The central government and the county governor tried to locate the site in Buan in 2003, even though most local residents opposed it. In particular, the county governor submitted application documents for locating the facility despite opposition from the local assembly. At that time, the local referendum law had not yet been enacted, but local residents undertook a referendum voluntarily based on the suggestion of the Buan Anti-Nuclear Coalition. During this time, Mayor Park was an executive director of the Beautiful Foundation, which was founded by him in 2002 and has led a new philanthropic culture in Korea. He was a chairman of the management committee for the Buan referendum. 72.4% of Buan residents (37,540 among 52,108 residents) voted, and among them 91.83% of voters expressed opposition to the establishment of the disposal site in Buan. Even though the Buan referendum was not a legally binding action, its political influence was strong: the plan to locate the radioactive waste disposal site at Buan was nullified. Mayor Park vividly experienced problems associated with the use of nuclear power, so he has been critical of it. Mayor Park has served as the President of the World Mayors Council on Climate Change since 2012. Seoul has played the role as secretariat for East Asia in the International Council for Local Environmental Initiatives (ICLEI). The World Congress of the ICLEI was held in Seoul in April 2015. This means that Mayor Park is deeply aware of and concerned about the issue of climate change.

Also, as a person who has worked in the civil society sector as a human rights lawyer and a founder of several civil organizations – for example, People's Solidarity for Participatory Democracy (1994), Beautiful Foundation (2002), Beautiful Store (2002), and Hope Institute (2006) – and as a civic activist and social designer by himself, Mayor Park has devoted himself to protecting citizens' rights and expanding citizen participation in decision-making and implementation processes. This aspect of Mayor Park's leadership is closely related to the way he promoted the OLNPP. In fact, governance entities were established based on his initiatives. He suggested how to organize governance entities and prepared the basis for their installation and operation by amending ordinances.

The benign political landscape and climate of Seoul also contributed to the advent and promotion of the OLNPP in Seoul, and promoted citizen participation. Mayor Park belongs to the Minjoo Party of Korea (formerly the Democratic United Party), in opposition nationally but a majority on the Seoul municipal council, which favours the democratic involvement of citizens in decision-making and implementation processes. Mayor Park was elected through a by-election in October 2011. The previous mayor, Oh Se-hoon, who belonged to the national governing party, was reelected in 2010 in a close race against opposition party candidate Han Myung-sook. However, Mayor Oh resigned his office after taking responsibility for the failure of a referendum concerning adoption of a free school meal plan for all students. Oh opposed the plan and called for a referendum to decide whether to give the free meal to all students or only to students from low income families. However, voter participation was only 25.7%, far below the minimum of 33.3% necessary to count the vote. Mayor Park ran for the mayoral by-election as an independent candidate backed by the opposition party, and then joined the opposition Democratic United party after getting elected.

In the 2010 local election, the opposition Democratic United party had gained many more seats than the ruling Grand National Party (79 vs. 27) on the city council. This made the political landscape and climate benign for Mayor Park to initiate the OLNPP. The city council was not a barrier in promoting the OLNPP in terms of budget allocation and amendment of the ordinance. In the local election undertaken on 4 June 2014, Mayor Park won another term; moreover the opposition Democratic Party also retained its majority status by winning 77 out of 106 seats on the city council, while the ruling Saenuri Party took the other 29 seats. This again is a favourable condition for the second phase of the OLNPP.

A noticeable movement from regional heads in February 2012, also created a friendly social climate. Forty-five (later 46 when Suncheon City joined in November, 2012) of the 226 district or county leaders announced a 'City Declaration for Post-Nuclear and Energy Transition' in order to create a substantial change in the energy policy of the local governments. Common action plans agreed upon by participating local leaders were: establishment of an energy ordinance, establishment and implementation of energy demand saving plans, expansion of renewable energy use through citizen-driven energy cooperatives, closing aged nuclear power plants and opposition to additional construction of new nuclear power plants, and a switch in national energy policy. Kim Sung-hwan, the leader of Nowon-gu, a residential district of Seoul, where radioactive materials were accidentally found in asphalt by a local resident was the first proposer. Fifteen out of the 25 district leaders in Seoul joined this movement. Mayor Park gave a keynote speech at the symposium for post-nuclear energy transition, which was held right after the city declaration event. He emphasized the necessity of moving toward a sustainable energy system and beyond nuclear power.

Most politicians who joined in the 'City Declaration for Post-Nuclear and Energy Transition', as well as Mayor Park, are experiencing a metamorphosis following nuclear-related incidents, especially the Fukushima nuclear disaster, and they emphasized the importance of citizen involvement. As decision-makers, they have been pursuing policies for energy transition as much as they can according to the authority given to local governors and district leaders.

3 Transparent information sharing and interactive communication with citizens

Interactive bilateral communication, not one-sided information delivery, is essential to expanding participation by citizens. The SMG has tried various experiments. It has tried to ask for and listen to the opinions of the citizens. Communication methods have been designed to encourage citizens to participate in the OLNPP. The city government coined a new concept called 'Policy Listening Workshop'. In general, this kind of event was just a policy workshop; however, the city government expressed its attitude to the citizens through the word 'listening'.

The city government established the Energy Information Center on the first floor of the City Hall's Seosomun Annex to deliver information about the OLNPP, to encourage the participation of citizens, and to offer consulting services for energy saving. It opened a webpage about the OLNPP (http://energy.seoul.go.kr/), where all relevant information can be shared and questions raised by citizens can be answered. Citizens can compare their district's electricity consumption with that of another district. There is also a section in which citizens can suggest ideas to promote the OLNPP. The city government holds public competitions more than 100 times a year to boost the participation of citizens and to make their ideas realized in daily life.

The name of the initiative, One Less Nuclear Power Plant, is powerful and strategic. Instead of presenting the city's targets to save energy or to reduce GHG emissions in measurement units, such as 2 Million TOE, the name is very stark and direct, which makes it easier for the citizens to understand. The city government invented a character, a brand identity (BI), and posters for easy and intimate communication as shown in Figure 4.5. The character on the poster is sunlight angel Donggri (meaning round shape in Korean), who symbolizes the sun. The city government has published many manuals to deliver guidelines for projects under the OLNPP by utilizing the Donggri character. Messages on posters were designed to be delivered by highly respected historical characters such as Admiral Yi Sun-sin and Sejong the Great.[4] BIs can also be seen on the left side of Figure 4.5. Posters, as shown in Figure 4.5, were also made to convey the main aims, values, and targets of the OLNPP. These posters were placed where visibility and accessibility to citizens are high. These images were donated by famous artists who were willing to be a part of the OLNPP. Brochures, leaflets, and newsletters repeatedly used these characters and BIs to approach citizens in a more friendly manner.

The Seoul Institute, affiliated with the city of Seoul, has created energy-related information graphics. Visualization of information on energy consumption enables citizens to recognize the current status at a glance. This is a more powerful way to stimulate citizens to think about the problems they are facing.

In summary, the city government communicates with citizens in a softer and more intimate manner, while emphasizing the importance of the participation and cooperation of citizens, in order to successfully implement the OLNPP. The slogan for the second phase of the OLNPP is 'Making an Energy Self-Sufficient City Together with Citizens'. In the evaluation report for the first phase of the OLNPP, it was described that the whole performance of the initiative was done through the citizens: the citizens were the main characters: 'The citizens created, the citizens reduced, and the citizens were the energy itself!'

4 Experts and NGO activists as policy advisors and facilitators

One of the bases for successful energy governance is the existence of mature civil activists and experts who have pursued an alternative path of sustainable energy. In

Figure 4.5 BIs and posters of the OLNPPI

Note: 1) OLNPP BIs; 2) Seoul, 'City of Sunlight'; 3) OLNPP, Seoul's energy policy for future generations; 4) Let's reduce CO_2 emissions and save the Earth; 5) Let's CO_2 reduce emissions and clean the whole universe

Source: the OLNPPI homepage, http://energy.seoul.go.kr/

spite of the not-so-short history of anti-nuclear (currently post-nuclear) movements and energy transition movements, there have not been enough opportunities to experiment with alternative energy options in South Korea (Yun 2015). It is the SMG that has provided such space and opportunity for alternative options.

Seoul's successful achievement of the OLNPP and successful cooperation with citizens are rooted in the voluntary actions taken by nonprofessional citizens. Although the institutional arrangements of the city government should not be discounted, such grassroots action has also been driven by growing exposure to the consequences of energy addiction and climate change. In the wake of Fukushima, the citizens of Seoul experienced rolling blackouts on 15 September 2011, witnessed the damage from heavy rain perceived as being linked with climate change, and became more aware of social conflicts over the construction of high-voltage transmission lines to support demand for electricity in the Seoul Capital Area. These events worked as emancipatory catastrophism and inspired metamorphosis, leading to the growth of energy citizenship and the energy-saving and energy-producing behaviour changes that followed.

The deliberative energy governance that the city government has pursued is an effective element for the successful promotion of the OLNPP. Even though local governments have not been given strong authority to exercise influence in the field of energy policy, the city of Seoul has expanded the areas it can engage in within that limited authority and has conducted this grand energy transition experiment by utilizing deliberative energy governance. It also mobilized expertise and invited lay citizens who are aware of the risks associated with using nuclear energy and climate change. Numerous ideas could be shared and social networks and solidarity could be strengthened. Participation in the OLNPP promotion process was also a learning process that contributed to capacity building. More particularly, it was deliberative governance, in which participants could express their ideas and persuade each other through discussion. This process could be a strong basis for social consensus building.

5 Educational programmes

Educational programmes played an important role in stimulating citizen participation and energy practices. The field of energy is not easy for lay citizens to deal with because it includes relatively complicated concepts and sometimes requires comprehensive understanding. Under the hard energy system, energy was always given to citizens, usually at affordable prices in South Korea. Energy, in particular electricity, was taken for granted in citizens' daily life. Thus, people did not ask where electricity was generated and transmitted. The four major events aforementioned revealed the importance of energy in citizens' daily life, but learning was required to understand these complicated concepts and issues.

Those who recognized the risk associated with nuclear power and climate change, for example residents of Seongdaegol in Seoul, were eager to know more detailed information and invited experts to their village. Experts and NGO activists organized post-nuclear and energy transition schools and educated citizens to be aware about the importance of energy and self-practice in daily life. SMG supported energy self-sufficient villages that held energy education programmes.

Usually, energy education is regarded as an essential factor for energy practice, or prerequisite of energy citizenship, because if citizens understand what causes problems and how to solve them, they will take action to solve them. There is an interesting finding, however. Through in-depth interviews, Jo and Yun (2016) found that energy practice stimulates the need for learning. If there is an existing community with good networks among members, some members will make efforts to save energy along with others despite having insufficient knowledge themselves. Through community activities and their co-practice to save energy, some members gradually acquire the identity of energy citizens and try to learn more about energy issues.

5 Conclusion: challenges and barriers lying ahead

Seoul has challenged the conventional fossil-fuel and nuclear-based centralized energy system. Instead, it pursues a safe, environmentally friendly, and ethical energy system, which is a decentralized distributive energy system based on the conservation of energy, the improvement of energy efficiency, and expansion of renewable energy use. In that Seoul is the capital of South Korea and one of the world's megacities, its challenge and experiment deserves great attention. The grand experiment Seoul has conducted is still on-going.

Even though SMG has achieved the target of the first phase of the OLNPP through active citizen participation supported by political leadership and expert engagement, there remains a long way to go. The success Seoul has accomplished so far does not guarantee future success. As a local government, room for action Seoul can utilize is limited, but also that Seoul can demonstrate an alternative method. Thus, it is important to evaluate the shortcomings of the past performance.

Even though deliberative energy governance has been pursued, there are many things that need to be improved. The Citizens Council, consisting of various leaders from the private sector, has held meetings only twice, unlike the Implementation Council. In order to have substantial participation among civil society, the Citizens Council has to be actively involved. In addition, although the Policy Listening Workshop and other forums were held, the range of participants was still limited. Since the number of participants in those events is physically limited, more flexible forms for such events need to be considered.

According to a survey conducted by SMG, targeting an on-line panel of 2,151 residents between 12 and 14 March 2014, the rate of awareness of the OLNPP was 71.4%. 'Know well' accounted for 14.3% and 'know roughly' accounted for 57.1%. Concerning the media through which citizens are informed about the OLNPP, internet portal sites showed the biggest share (32.7%), followed by promotional materials issued by the city of Seoul (27.9%), articles in newspapers and journals (27.1%), city and district home pages (26.7%), and broadcast stations and cable channels (21.9%). However, with regard to the level of awareness about the OLNPP in a survey conducted by the Seoul Institute, the rate of 'know' was just 36.5% ('know well' 7.3% and 'know a little' 29.2%). Those who did not know about the OLNPP comprised 28.6% in the online survey done by Seoul city and 36.6% in the survey done by the Seoul Institute. This implies that the level of awareness is not very high. Improving the rate of participation and practice of more citizens is required.

The low level of citizen awareness about the OLNPP is closely related to the exclusion of major conservative mass media. Since Mayor Park stands on the progressive side and belongs to the Minjoo party of Korea, major influential conservative newspapers, including the *Chosun Daily* newspaper and the *Jungang Daily* newspaper, and major broadcasting stations such as KBS and MBC, did not report about the OLNPP frequently enough to draw substantial attention from the citizens. For example, although the Seoul International Energy Conference deserved massive media coverage, major media did not report the event. This unfavorable media environment was a serious barrier to active promotion of the OLNPP 2. More effective communication strategies must be developed. For the further promotion of the OLNPP and the expansion of similar initiatives in other local governments, national energy policies and institutions have to be changed. However, Seoul cannot wait for changes to occur at the national level. It has to prove that its policy direction is the right way to go. More concern and participation from citizens are also necessary to achieve the goal. Smarter deliberative energy governance must be implemented.

The most recent opinion poll, in December 2015, showed a slightly recovered support for nuclear power. 85.1% of respondents agreed to the necessity of nuclear power in Korea, even though only 41.0% of respondents thought it was safe. It can be inferred from this data that Koreans view nuclear power as a necessary evil. Nevertheless, more people supported just maintaining current nuclear capacity (40.2%) than expanding it (33.7%). The share of respondents who supported reduction of nuclear capacity remained the smallest, at 21.1%, but this is much higher than the 9.3% in 2010, before the Fukushima nuclear disaster. These survey results imply that many Korean citizens reluctantly accept the necessity of nuclear power because they think there is no alternative. Even in South Korea, the success story of Seoul has not been disseminated well nationwide. This model case has to be circulated within the Korean society and, furthermore, has to be transferred to other countries.

Notes

1 Distrust in nuclear safety was deepened because of disclosure of consecutive corruption scandals after the OLNPP started. In November 2012, another corruption and bribery scandal was revealed. Some of the reactors were found to have substandard parts supplied via fabricated quality certificates. More shockingly, in May 2013, inspectors discovered that some reactors operated components with fabricated test certificates, some of which were directly related to safety. Due to these scandals, three reactors were suspended, which led to risks of power shortages. The Korean public soon became very suspicious about nuclear safety. Results from the opinion poll conducted in May 2014, right before the second phase of the OLNPP started, show decreases in public acceptance and increases in apprehension about nuclear safety. This trend is deeply connected

with the corruption and bribery beyond the Fukushima nuclear disaster. 75.3% of the respondents agreed with the necessity of nuclear power generation. Although this percentage remains high compared to other countries, it has decreased from 89.4% in 2010 and 78.3% in 2011. With regard to the number of nuclear power plants, 37.3% agreed to increasing them, 31.4% agreed with maintaining the current level, and 27.3% agreed to decreasing nuclear power plants. Only 29.8% of the respondents considered nuclear power to be safe and 31.5% showed willingness to locate nuclear power plants within their communities.

2 Subway stations are one of main objects for LED light replacement. In the first phase, a total of 430,000 lights at the subway stations were replaced with LED lamps by May 2014. The second phase aims to replace a total of 220,000 lights in all of its subway cars with LED lamps as soon as possible.

3 SMG has been awarded several prizes. In June 2013, the '2013 UN Public Service Award' was given to the Eco-Mileage System, one of the representative projects of the OLNPP. Expansion of a culture of energy conservation and reduction in energy consumption through civic participation were highly praised. In November 2013, the Climate Action Leadership Award was given to the OLNPP by the World Green Building Council (WGBC) for its comprehensive campaign to reduce energy consumption by buildings through implementation of BRP and expansion of new and renewable energy use. In April 2014, SMG was awarded the 2014 Earth Hour City Challenge (EHCC) by the World Wide Fund for Nature (WWF) and Local Governments for Sustainability (ICLEI) for its commitment to combating climate change and efforts to reduce its CO_2 emissions along with solving global energy and environmental issues.

4 Yi Sun-sin (28 April 1545 – 16 December 1598) was the most admired Korean naval commander. He is very famous for his victories against the Japanese navy during the Japanese invasion of Korea, the Imjin War. He is shown on the third poster from the left, in which he says, 'Let's reduce CO_2 emissions and save the Earth!' Sejong the Great is the most respected king in Korea. He was the fourth king of the Joseon-dynasty Korea. He accomplished too many great performances to describe, including the creation of Hangul. On the last poster, he encourages people 'Let's reduce CO_2 emissions and make the whole universe clean!' These posters attracted a great deal of citizen attention.

References

Beck, Ulrich. (2015). Emancipatory catastrophism: What does it mean to climate change and risk society? *Current Sociology*, 63(1), 75–88.

Beck, Ulrich. (1992). *Risk society: Towards a new modernity*. SAGE Publications.

Devine-Wright, P. (2007). Energy citizenship: psychological aspects of evolution in sustainable energy technologies. In J. Murphy (ed.) *Governing technology for sustainability*, 63–86. London: Earthscan.

IEA. (2015). Energy and climate change. *World Energy Outlook Special Report*.

IPCC. (2014). *Climate Change 2014: Synthesis Report*.

Jo, Miseong and Sun-Jin Yun. (2016). A shift of viewpoints on the formation of ecological citizenship through the process of energy transition movements: Based on a qualitative case study on an energy self-sufficient village in Gwanak-gu. Under review process.

Korea Energy Economics Institute. (2015). *The Year Book of Energy Statistics 2015*.

Lee, Soo-Jin and Sun-Jin Yun. (2013). The reporting coverage and frames of the electricity crisis in four major Korean daily newspapers. *Journalism & Communication*, 17 (4), 175–208. (in Korean)

Lee, Hwa-yeon and Sun-Jin Yun. (2013). An analysis of news coverage on conflicts concerning transmission line construction in Miryang: From a perspective of environmental justice. *Economy & Society*, 98, 40–76. (in Korean)

Lovins, Amory. (1976). Energy strategy: The road not taken?, *Not Man Apart*, 4(20), 5–15.

Park, Jongmun and Sun-Jin Yun. (2016). The formation process of energy citizenship in urban community energy transition movements see through the case of SUNGDAEGOL village in Seoul. *Space & Society*, 26(1), 79–138. (in Korean)

Seoul Metropolitan Government. (2013). *One Less Nuclear Power Plant 2012.*

Seoul Metropolitan Government. (2014). *One Less Nuclear Power Plant, Phase 2.*

Yun, Sun-Jin. (2015). Korea's nuclear policy: Past, present, future. *KAS Journal on Contemporary Korean Affairs: Environmental Policy in South Korea – Problems and Perspectives*, 117–170.

5 Energy democratization in Taiwan

Current problems and the civil society's advocacy

Shu-Fen Kao and Ying-Feng Chen

I Introduction

Although climate change has become a real threat for global communities, the growing demand for fossil fuels has led many countries towards alternative forms of energy. Thus, nuclear power has experienced a resurgence in recent years, as it is viewed as a viable 'low-carbon alternative'.[1] However, the devastating accident that occurred when three nuclear reactors were disabled after a tsunami hit the northeast coast of Japan in 2011 have raised concerns about the safety of nuclear power. In Taiwan, the Fukushima accident has had a massive impact on public's perception of nuclear power, (Ho et al. 2014; Liang and Lee 2014), which has led to a series of anti-nuclear power actions, including advocacy that halted the construction of a new nuclear plant in New Taipei City. Thus, as nuclear energy is no longer an option, due to public disapproval, Taiwan's government has instead started to develop other forms of energy. For instance, beginning in 2009, the government not only initiated a series of renewable energy proposals, but has also introduced the FIT (Feed in Tariff) system, which is based on EU practices and policies.

Moreover, in 2009, Taiwan's government passed the 'Renewable Energy Promotion Act' (REPA afterwards), which embodied an incentive mechanism for the development of solar energy by granting FIT benefits for 20 years to projects involving the expansion of renewable energy. Solar energy has great potential and may be a viable option in Taiwan, as southern Taiwan has the sunny climate needed for widespread implementation of solar power. However, as of 2015, less than 2% of Taiwan's electric power generation was from renewable energy sources. What institutional and structural failures account for the lack of transition to renewable energy in Taiwan? How has civil society responded to various barriers and obstacles preventing widespread implementation of renewable energy while advocating for energy democratization in Taiwan? In order to answer these two questions, the authors first present a review of the literature concerning energy transition and energy democratization as an analytical base. They then consider how the authoritarian and state-centred legacies of the developmental state have posed challenges to Taiwan's energy transition. In the fourth section, the authors propose how civil society has advocated for energy democratization in recent years. Lastly, the authors offer some suggestions for strategic actions to accelerate energy transition in Taiwan.

II Energy transition and energy democratization

There is a fundamentally distinctive agenda in the transition from fossil fuels to renewable energy, which differs from previous energy transitions. Today, civil engagement and local participation are both imperative and requisite. In fact, for several reasons, public engagement, decentralization, and local diversity are key factors in renewable energy discourse and development.

First, the world's energy structure, which has long been based on traditional sources of fuel, reflects the unequal distribution of power in global politics. The dependency of the whole world, limited by geological restraints, dwindling supplies of fossil fuels, and the uncertain prospects of widespread nuclear power have shaped a privileged class comprised of politicians and corporate interests. Moreover, in order to sustain their advantageous position, this dominant group, spawned from the resource monopoly, seeks to suppress and quell the voice of the people. While renewable energy, in its various forms, appears accessible to every local community, the transition to renewable energy may likely reshape social networks and create a more egalitarian power structure, based on diverse energy sources, which illustrates the increasing empowerment of civil society (Rifkin 2013). In this sense, this process of migrating to renewable sources of energy, in concert with further democratization and equality, has emancipated society from energy deprivation and power abuse.

Large-scale renewable energy projects patronized by national governments, as seen in both the United Kingdom and the Netherlands, often culminate in bureaucratic rigidity and unsatisfactory power efficiency (Kemp and Loorbach 2006). Thus, the participation of ordinary citizens, local governments and neighbourhood communities in democratically influencing power policies and infrastructures have been essential in not only promoting and supporting innovations in renewable energy, but also in breaking the monopoly of energy politics. In addition to increased implementation and application of renewal energy sources, which is a result of advocacy by civil society, the social awareness and technological innovations inspired by the process were also the keys to the successful transition to renewable energy.

Second, the transition to renewable energy has been replete with vast uncertainties in terms of technology as well as various social risks. Addressing the technological problems alone has proven to be extremely difficult for many national governments and corporate innovators. As seen in several EU countries, great problems occurred when governments and business interests within the energy sector set the agenda for the transition to renewable energy. In fact, given the high fixed costs incurred within the energy sector, it is unlikely that radical changes would emerge from power generating institutions. The agenda for any transition to new sources of energy must inevitably come from society. Thus, a unifying social platform must be provided in order to establish social connection and create common ground (Frantzeskaki et al. 2012).

Energy security and stable energy prices are always top national priorities. Therefore a government may be reluctant to risk a transition to alternative sources of energy without first investigating and understanding the feasibility of these new supplies of power and their related technology. In fact, despite the existence of environmental monitoring mechanisms in countries like Denmark and Germany, great modifications are still slow to come in terms of energy innovation. Again, it has only been through local communities and NGOs that alternative sources of energy have been deemed both feasible and possible (Lund 2014).

Therefore, in the long run, local communities, discussion forums, citizen fund-raising mechanisms, and social groups play vital roles as intermediaries, coordinating support for different projects amidst various challenging social and political circumstances. In fact, with the cushion created by tight-knit social networks, several radical ‘niche’ innovations have become practical realities (Hargreaves et al. 2013).

Third, the diverse features of renewable energy sources preclude the command structure from simply focusing on a single component, as even this requires trans-sector thinking and collaboration. A centralized, top-down system, promoted by technocracy, may fail to recognize the complexity of energy transition and often forsakes linkages with civil society, and thus is more inclined to only focus on short-term issues (Hess and Mai 2015). Therefore, the inclusion of multiple government departments, the business sector, and civil society is pivotal in the energy transition process (Adamsa et al. 2011; Seyfang et al. 2013).

Germany’s successful transition to renewable energy was mainly due to decentralization policies and diversified energy innovations, which contrasts greatly with other European countries that cling to large, centralized systems of power generation (Szulecki et al. 2015). In particular, the inclusive planning culture integrating diverse stakeholders inspired the German government to make binding decisions that facilitated more rapid energy transition (Schweizer et al. 2014).

As empirical evidence in the US and EU shows, a distinct regulative culture exists between both the state and private interests.[2] Moreover, the evidence also shows that the best way to reconcile conflicts between energy efficiency and social welfare is to maintain local diversity of energy innovations, which may be ensured by open-minded, honest, and transparent arrangements in the energy transition process (Khan et al. 2014).

III Taiwan’s state-centred developmentalism

The unsatisfactory development of renewable energy in Taiwan might seem irrelevant from a technological or economic standpoint. Instead, the problem may be linked with the country’s legacy of statist developmentalism. Taiwan’s government creates markets, but only nurtures certain sectors of the economy, which has directly affected the trajectory of energy transition. Thus, Taiwan’s traditional development strategy only focuses on a limited number of strategical sectors, normally in capital and energy-intensive areas, and grants privileges to accommodate them (Johnson 1987; Wade 2003). Although Taiwan has experienced a democratic evolution since the 1990s, the authoritarian and state-centred legacy of the developmental state still poses several challenges to Taiwan’s energy transition in the following dimensions.

1 The energy monopoly of state-owned enterprises (SOE)

In order to guarantee a stable power supply and meet national energy goals, Taiwan's government has allowed the Taiwan Power Company (TP afterwards) to operate with a virtual monopoly over the energy sector for several decades. In fact, in spite of all the democratic changes and policies that have developed since the 1990s, the TP is still the only agency authorized to generate and sell electricity in Taiwan. Every kilowatt of electricity, including those generated by private plants, may only be distributed through the TP power grid. Thus, this monopoly has led to great rigidity and inflexibility in the energy sector, which is discernible in several areas.

First, TP lacks incentives to bolster renewables. According to the REPA, TP was directed to buy renewable energy through the FIT system. In other words, the greater the amount of electricity generated by renewable energy sectors, the more TP had to spend. This zero-sum game left TP with no impetus whatsoever to facilitate renewables or even to make purchases easier, as there was hardly reciprocity between the company and the FIT providers.[3] This intransigence has also resulted in limited information flows to local communities, as most people receive little, if any information about citizen-based renewable energy programmes.

Second, although TP is a state-owned enterprise, the cost of power generation does not reflect market trends. While 98% of Taiwan's energy is imported, the price of energy has been surprisingly cheap, given its scarcity. In fact, based on Taiwan's state-socialism legacy, the price of energy, which is a vital resource and a means of promoting market competitiveness, is strictly controlled and subsidized by the government.

Since cheap power may also be linked with the provision of social welfare, Taiwan's government is reluctant to raise energy prices or collect carbon taxes, as this is not only politically inexpedient, but also threatens the legitimacy of the developmental state (Yang 2013). In fact, in 2014, the price of energy for household use in Taiwan was the third lowest in the world, while the price for industrial use was the fourth lowest.[4] This subsidization of Taiwan's energy sector has twisted free market mechanisms and weakened the incentives for society to seriously consider energy transition.

Third, as TP is a state-owned enterprise that monopolizes energy sectors. The lacking of competition has led to inefficiency and complacency. In addition, as previously demonstrated, the FIT system is a costly burden. Not only does TP have to purchase electricity from FIT provides, but it also has to cover the costs of increasing and upgrading the power grid infrastructure. Moreover, as the FIT mechanism is by no means self-enforcing, TP's voice has also been marginalized.

The inefficiency issue is also plagued by administrative red tape. For technical reasons, any power grid infrastructure installation must be locally customized, which has inevitably led to a fragmented structure, as power companies have to respond to geographical heterogeneity.[5] In addition, this fragmented structure, combined with the fact that TP is a state-owned enterprise, has caused chaos and inefficiency in the process of reviewing renewable power projects. Many applications are turned down, for a wide range of reasons, and most companies and individuals are discouraged from filing grid mergers. As a result, many people have lost enthusiasm for renewable energy projects and challenging big power corporations that aim to exploit the FIT system.

The FIT system has been proven to be successful in several countries. However, while the supply issue has been addressed to a certain extent, FIT has hardly facilitated better energy efficiency on the demand side under Taiwan's energy price structure. Ironically, in spite of the government's great efforts to reduce consumption and patronize renewable energy through FIT, demand for power continues to grow, as local participants exploit the cheap energy provided by TP while selling back the power they generate at higher FIT rates.[6]

The artificially low cost of energy, which is sustained by governmental subsidy, has created a great level of injustice between power producers and consumers, as well as taxpayers and manufacturers. More importantly, this subsidy has created a direct challenge to economic growth, as it not only inhibits a structural reform of the energy price system, but also prevents tax cuts, which has been a primary policy measure that has long contributed to the success of Taiwan's developmental state.

2 The technocracy-based knowledge system and policy-making

While energy matters are directly linked with national security, many governments worldwide are more inclined to adopt conservative strategies in terms of power issues (Grundig 2006). Moreover, due to the high level of uncertainty during an energy conversion period, a 'trade-off' discourse may compromise society's confidence in implementing a transition project (Pralle and Boscarino 2011). Additionally, incumbent technocracies, existing energy sectors, and scholars who may benefit from the current structure may seek to construct epistemic obstacles, through knowledge statements, as these groups of stakeholders have agency, and are thus motivated to mobilize and resist reforms (Hess 2014). Thus, it has been difficult for many societies to overcome these institutional and epistemic barriers. Yet, in a developmental state like Taiwan, which is highly dependent on a technocratic-planned economy, it is even more challenging to hold discourse on renewable energy.

Taiwan's developmental state has successfully nurtured the petrochemical and semi-conductor industries, and has thus become tightly entangled with these energy-consuming sectors (Chu 1994). Yet, a key component of this growth model is a consistent supply of cheap energy from the state. This has stymied energy reform, as a rapid shift to renewable energy would cause vast uncertainty and could have a negative effect on business chains under state control. In fact, resistance from a corporate regime bound to the coal, gas, and nuclear production sectors has intentionally negated the benefits of increased renewable energy development in the UK (Geels 2014).

Moreover, the dominance of technocracy in knowledge production and the political arena has propagated the idea that energy transition is risky and dangerous, which has in turn prevented reforms and substantial changes, such as closing nuclear power plants.[7] While community-based power generation is highly associated with trust and peer impacts,[8] many energy initiatives have been marginalized or even ignored, as civil society has been unable to advocate for change. Thus, centralized developmental statism has guided economic growth and interceded in the market.

Additionally, path-dependent policy-making models have deprived the technocracy from understanding the depth of vital and comprehensive social upgrades. Moreover, the technocracy, failing to recognize the social dimension, has simply treated the energy transition as a technological issue. Therefore, the government has passed on the responsibility for promoting renewables to TP and a state-owned lab agency, Industrial Technology Research Institute (ITRI), which provides services such as technology transfer and industrial innovation.

This centralized technocracy, owing to its technological and under-ranked status, has also led to the impotence of social advocacy. Despite their status as a research institution, ITRI seems incapable, both financially and in terms of staff, of guiding a nationwide energy transition.[9] In fact, without assistance from local administration, much of what ITRI has accomplished has centred on resolving investment obstacles for the business sector, rather than advocating for renewable energy. Without good information, and without confident leadership, it is not surprising that a 'techno-group' would be unable to push for energy initiatives at the local level. Consequentially, the optimal option for ITRI has been to push for more grand-scale mega projects.

3 The export-oriented mindset of economic growth

Following tradition, Taiwan's state-guided developmentalists constantly seek out the next competitive sector to nurture. Due to Taiwan's small market size, export-oriented developmentalism, which operates through the strategic selection of potentially successful business sectors, has also limited energy transition. Rather than treating energy transition as an opportunity for growth and the realization of a more sustainable economic model, the government sees renewable energy as a 'commodity' for export. For example, Taiwan is second only to China in the production of solar panels. Yet, despite the high level of solar technology in Taiwan, the implementation of solar energy island-wide remains limited. Surprisingly, although 95% of the solar energy generated in Taiwan is exported, it only represents less than 1% of the total amount of energy generated.[10]

The process to produce solar panels is not only carbon-intensive, but also, the export-oriented manufacturing of solar panels is detrimental to the sustainability of Taiwan's energy structure. In fact, since solar panels are only produced for export, there is neither need nor room to seek collaboration with local communities and civil society to facilitate renewable energy agendas.

In sum, long-term centralized developmentalism has made the Taiwanese government more reluctant to embrace a more dispersed, equal, and citizen-based power generation system. Therefore, the social innovation of renewables, as a flag of energy democratization, thus poses a direct challenge to the black box sealed by technocracy and the existing energy-intensive sectors.

IV Civil society's advocacy for energy democratization

As discussed in the third section, the authoritarian and state-centred legacy of the developmental state poses several challenges to Taiwan's energy democratization. However, civil society has been trying to break these constraints through alternative energy option discourse, strengthening local energy governance capacity and direct action to construct citizens' power plants to achieve energy democratization.

1 Alternative energy option discourse

In the aftermath of the Fukushima disaster, Taiwanese society started to focus on nuclear safety issues. Many posed questions such as: Will Taiwan have power shortages without nuclear energy? What alternative energy resources could Taiwan utilize in order to secure a stable supply of electricity that is neither detrimental to public health nor harmful to economic development and quality of life?

In response to these types of questions, Taiwan's government, civil society, and various scholars have introduced several programmes and plans for the country's future energy needs. For instance, Taiwan's Bureau of Energy, which is part of the Ministry of Economic Affairs (MOEA), proposed an Energy Policy Program to set forth Taiwan's national energy structures for 2011–2020.[11] Since nuclear energy was a key component of this programme, it is not surprising that many began to call for the construction of a fourth nuclear power plant in Taiwan. While the government seeks to maintain national economic development and attempts to reduce carbon emissions, ostensibly through the use of nuclear power, the public has become increasingly uneasy. The Fukushima disaster has provided many lessons for Taiwan, both in terms of nuclear power plant deployment as well as the inherent and ongoing risks presented by nuclear power. Civil society groups have challenged the government's pro-nuclear power claims and have even introduced several discourses of their own in regard to nuclear power.

1.1 Debunking the claim of nuclear safety

During the debate over the controversial fourth nuclear power plant,[12] Taiwan's nuclear regulator, the Atomic Energy Council (AEC), indicated that the country's three existing nuclear plants had been built to withstand earthquakes of magnitude 7 and above and tsunamis of 12 to 15 metres. The AEC even used metaphors, stating that the fourth nuclear power plant was to be 'built on a rock', which was 'as solid as the Buddha sitting on the lotus throne', in order to persuade the public that nuclear power was definitely safe.[13] But experts involved in the relief work following the 921 earthquake, where 2415 people died in 1999, cast doubts on the AEC's safety claims. These experts argued that when seismic plates move, they may fragment or shatter rock, which releases a great deal of energy. Therefore, governmental officials should be more sceptical about nuclear safety, instead of boasting about it.[14] As one environmental NGO leader also emphasized, 'the seismic design of the first nuclear plant is 0.3g, and the second, the third and the fourth nuclear power plants are 0.4g, which are far worse than 0.6g, the original Japanese nuclear power plant seismic design. Responding to the earthquakes, the Japanese nuclear power plants have begun seismic intensity upgraded to 1.0g. But Taiwan is still ignoring the threat of earthquakes'.[15]

1.2 Debates on energy structure, power shortage and energy efficiency

Another debate concerns the question of shutting down the three existing nuclear power plants in Taiwan, and if this will lead to power shortages. Environmentalists claim that Taiwan will not suffer from a lack of electrical power if the existing three nuclear plants are closed, but TP and the MOEA refute this claim.[16] According to statistics[17] from the Bureau of Energy, nuclear power comprised 17.2% of the total amount of power generated in Taiwan in 2010. The MOEA keeps emphasizing that once the nuclear facilities are shut down, Taiwan's power reserve margin will fall to 10% immediately. Moreover, the MOEA also points out that Taiwan's power system is an isolated system confined to an island, which contrasts greatly with the European system that is based on a series of grids that support each other. The danger of an isolated system is that when spare capacity is low, there is a risk of power shortage or rationing if the electricity load suddenly increases or if unit failure occurs. The MOEA also stressed that renewable energy will only generate 4% of the power needed, which is far less than what nuclear plants can provide. In fact, the MOEA also claims that if Taiwan's three existing plants are closed, the reduced amount of electrical power will force between one-third to one-quarter of Taiwan's manufacturers to close.[18] Yet, environmental groups refute these claims, as one of the anti-nuclear movement leaders explains in an interview:

> 'The so-called reserve capacity refers to the current generating capacity of power generation systems. Compared to the highest consumption of electricity for a specific day annually, the power supply capacity required can be much higher. The statutory reserve capacity rate in Taiwan is 16%. In recent years the reserve capacity rate in Taiwan is more than 20%. According to the Bureau of Energy's 2010–2019 long-term load forecasting and power development plan, the net peak capacity at the end of 2010 was 44,827 MW. What does this mean? It means if we immediately shut down the operation of the six existing nuclear power units – in other words, deducting the 4,884 MW of net peak capacity generated by these six reactors – Taiwan still has nearly 40,000 MW of net peak capacity. If we stop using nuclear power, the reserve capacity rate would be reduced to about 11%, but this spare capacity is higher than that of Japan and South Korea. I think as our society needs to exercise an energy-saving lifestyle and keep the peak load at the minimum value, as it has been for the past five years. So, even without nuclear power, the reserve capacity rate can still be maintained above 18%'.[19]

Taiwan's electricity consumption has been steadily rising, and in 2010, it peaked, reaching the highest levels in the country's history.[20] The industrial sector represents the highest percentage of use at 52.26% of the total consumption. In contrast, the service sector (shops, restaurants, and department stores, supermarkets) used 20% of the total, households used 18%, and agriculture just 1%. The loss rate from energy sector generation represents about 8.09%.[21] One environmentalists indicated that 'from 2008 to 2009, power consumption decreased in Taiwan, as a result of both household energy-saving and economic turmoil. As long as Taiwan can keep its usage of electric power down to the average electricity consumption in these two years, we can save 122 billion kWh of electricity, compared to what was consumed in 2010.'[22] This environmentalism went on to comment that:

> '… in fact, the reason why the government needs to substantially increase the reserve capacity is to develop industries that are high energy-consumers and pollutants, such as the petrochemical industry. In the past, Taiwan counted on coal-fired power generation, but in recent years, facing carbon reduction pressures, the government has turned to nuclear power. In other words, why does the government favour nuclear energy? It is because the government promotes energy-consuming industries and tries to persuade the Taiwanese people that Taiwan needs it … such as in the name of increasing employment opportunities … etc. The Ministry of Economic Affairs even warns the public: "no petrochemical industry, no daily necessities". You know what? If we look from the end side of consumption, petrochemical products in Taiwan are mostly for export … Taiwan uses almost half of its current electricity consumption for export. Is it worth it to let the Taiwanese people bear the risk of nuclear disaster? … I think it's time to seriously consider re-structuring our industrial structure …'[23]

Contrasting views also exist between the government and those who are against nuclear energy in regard to enhancing energy efficiency and renewable energy. While environmental groups think Taiwan should increase its energy efficiency to 4%, one expert, commissioned by the MOEA, concluded that over the past twenty years Taiwan has only achieved less than 0.5% of energy efficiency, and that energy efficiency worldwide is less than 2%. In addition, the Bureau of Energy indicated that the goal of renewable energy development should take the following factors into account: domestic and international renewable energy technology, its cost, development potential/land use patterns, weather and other natural, geographical conditions.[24]

In addition, Lee (2011) believes that renewable energy is less stable than nuclear power. He thinks that the greatest challenge to the development of renewable energy is land space. In fact, as he continues, even if the number of solar panels were to be maximized, the amount of electricity generated would only be about 14.2% to 20.5% of what a fourth nuclear power plant would generate. Moreover, Lee adds, as of 2009, there was a total of 182 units of wind power in Taiwan. Even if the total installed capacity was to be maximized, wind power generation would only generate 53.8% of the fourth nuclear power plant's potential capacity.

Members of anti-nuclear NGOs counter that the government and members of the pro-nuclear camp have over-stated the efficiency of nuclear power. In fact, as one advocate claims, nuclear power generation is only about 30% efficient. In other words, only one-third of the energy generated is converted into electricity, while the other two-thirds must be cooled with seawater. In addition, the environmental costs of an additional nuclear power plant have been taken out of the equation.[25]

1.3 Rights to choose green power

Faced with the global challenge to reduce greenhouse gases, and geographically located in an earthquake-prone region, Taiwan must begin energy transition immediately. Increasing numbers of people in Taiwan are speaking out about their right to choose green power, and showing their willingness to pay more for electricity in exchange for green power generation. For example, 46.58% of those who responded to a national survey (valid random sample size: 2000) indicated that they would be willing to double their current electric bills in order to eliminate nuclear power in Taiwan (Yang, 2011). In addition, one civil society group, Crazy for Green Power Action Alliance (CGPAA), continues to disseminate the details behind Taiwan's historically unfriendly renewable energy development policy and urges the government to seriously re-tool the current energy structure, expand renewable energy development strategies, and abolish nuclear energy gradually. The CGPAA also proposes the 'Green Energy Pricing Scheme for Taiwan under the Framework of the Current Renewable Energy Development Ordinance and Electricity Law', while suggesting that a green power pricing system could be implemented immediately in Taiwan. Finally, the CGPAA also encourages people to download an 'application form for green power purchase' from the Internet, fill it in and then send it directly to TP.[26]

2 Strengthening local energy-saving governance capacity

The government's energy-saving policies have been characterized as top-down and centralized approaches, which have mainly been implemented by TP. Under Taiwan's energy price structure, the FIT only tackled issues on the supply side, but hardly facilitated better energy efficiency on the demand side. In spite of the government's great efforts in promoting renewable energy through FIT, demand for electricity kept increasing, as local participants found they could exploit the cheap power provided by TP while selling back the power they generated at the FIT rate. Therefore, energy-saving is rather critical in Taiwan, as the country's energy grid is stressed during peak consumption periods in summer months, particularly during the process of nuclear energy phase out.

In April, 2015 the Energy Bureau of the MOEA proposed a smart energy-saving programme, which involved an investment of 3 billion NT dollars to subsidize efforts in 19 cities and counties aiming to save electricity by 2%.[27] NGOs welcomed the government's intentions, but were also concerned about the lack of local energy conservation plans. Civil society groups,[28] which have actively been involved in energy transition in various cities and counties, came together to form the 'Energy Transition Promotion Alliance' (ETPA afterwards). ETPA aims to mobilize grassroots support, promote energy education, and urge progressive implementation of energy transition at both the local and central levels. Inspired by the success of Seoul's 'One Less Nuclear Power Plant', ETPA stressed the importance of public participation as a critical key for a city's energy-saving plan.

In response to the MOEA's smart energy proposal, ETPA conducted a complete appraisal of the energy-saving policies proposed by each city and county that participated in the new programme. ETPA's goal is that local governments may be encouraged to seriously consider implementing effective energy-saving policies. Five features that ETPA used for appraising the local energy-saving policies include: the establishment of an energy committee involving civil society, promotion of local energy autonomy ordinances, design of policy mechanisms to facilitate more input from civilian resources, a demonstrated understanding of local energy consumption patterns, and a detailed plan showing the amount of administrative resources invested. In addition, ETPA also looked for 'potential effectiveness in suppressing peak power consumption' and 'promotion of other forms of public welfare'. These characteristics would thus be implemented in order to encourage local governments to effectively reduce electricity during peak demand, as well as to consider fair distribution of energy and the needs of the disadvantaged.[29]

One ETPA member indicated that both the central and local governments neglect the significant factor of peak load consumption in their energy policies. Another member pointed out that demanding each sector to reduce their energy consumption by 2% cannot solve the problem. Still, it is a fundamental challenge to identify the causes of high energy consumption in each city or county and take practical actions to reduce the consumption.

There are other structural barriers hindering the Energy Bureau's programme. For example, the Bureau has little authority or clout to coordinate with other government departments to tackle related barriers, such as regulations for the energy efficiency of new buildings, or incentives for banks to provide low-interest loans. In addition, the programme lacks urgency, as there are no mandatory plans. Thus, many local governments may not even participate, as seen with Hualien County's decision to opt out of the programme.

Moreover, there are a number of deficiencies in the planning of the smart energy-saving programme. For instance, the official evaluation criteria are one-dimensional and rigid, ignoring the key component of suppressing peak power consumption as a part of the overall energy policy. Also, the effects of the central regulatory system on the energy transition has not been fully examined. In addition, TP has passively provided information concerning local energy consumption, which stymies effective local energy planning. ETPA emphasizes that it's essential for local governments and the general public to realize what contributes to energy-wasting and how everyone can work to save energy. Furthermore, it's ironic that building insulation is excluded from the evaluation criteria for energy saving. Factories, which are probably the largest energy consumers, have also been excluded from the evaluation.[30]

In spite of problems discussed above, several local governments have actually made great strides in terms of energy saving. New Taipei City won first prize in an energy competition. In addition, the wisdom of experts and local citizens were combined to establish the 'New Taipei City Smart Energy-Saving Promotion Committee', which created seven topic groups aimed at integrating citizens, business interests, schools, institutions, and civil society, with the goal of promoting energy-saving. The New Taipei City government also selected two administrative districts, combined with local universities and non-profit organizations, to promote a participatory budget discussion with an energy-saving theme. This discussion allows for local people to have a discourse on energy saving, develop energy-saving programmes, and establish different local energy management patterns. In fact, ETPA considers New Taipei City as a model of local energy transition, due to its creative and participatory budget discussion, as well as comprehensive thinking in its policy formulation, cross-office integration, civil participation, and long-term planning.[31]

3 Constructing citizens' power plants

In addition to their public proclamations against nuclear power, Taiwan's NGOs have taken another step, which is to construct citizens' power plants in several communities. This will not only enhance the percentage of renewable energy being used in Taiwan, but also show that 'anti-nuclear' is more than just a slogan.

3.1 The 'one person, one kilowatt' social enterprise

HAND Initiative e.V., a joint Taiwan–Germany international organization established in 2002, seeks to promote the value of labouring and establish an environmentally friendly society. The main HAND office in Taiwan is located on the second floor of the Solar Library and Solar Energy-Saving Exhibition Hall at Taipei Youth Park. This building, called the Solar Leo House, combines various energy saving methods. In fact, the innovative combination of energy-efficient windows, state-of-the-art insulation, 'intelligent' lighting, and cutting-edge ventilation techniques used in building Solar Leo House have set new standards in the construction of new public buildings. In addition to these features, this pioneering building also has a remarkably thick front door, double-coated window panes and fewer windows facing north, which are all measures aimed at reducing heat exchange and protecting the environment.[32]

Instead of simply 'talking about' renewable energy, HAND Initiative e.V. has initiated a new civic movement, 'one person, one kilowatt social enterprise'[33] to build 'citizen's power plants' in varied regions in Taiwan and promote civic energy education and energy independence. The idea of 'one person, one kilowatt' is to utilize the power and resources of the people, through social enterprise and the co-reciprocal method, to provide alternatives to coal-fired and nuclear power generation. In addition, HAND's strategy is to seek partners to participate in the 'Sun Shareholders' programme. Each 'shareholder' is asked to contribute NT $100,000, in order to purchase photovoltaic panels. Individuals throughout Taiwan will rent out roof space to HAND, which will install the solar panels. As electrical power is generated through the panels, HAND can sell the electricity to TP and let the Sun Shareholders share the green benefit.

Unlike other large-scale solar photovoltaic instalment projects, which are largely located in rural areas with little visibility, the photovoltaic panels in the 'one person, one kilowatt' programme are used in communities throughout Taiwan, so they are highly visible. In turn, the high visibility of solar panels also aids on educating people about the value of renewable energy. Moreover, the people who rent out roof space for solar panels have a much more practical and concrete example of renewable energy. Residents in the community would also have more exposure to renewable energy, which in turn may lead to further knowledge and interest in green energy. Thus, a community-based approach to solar photovoltaic power generation is more likely to involve citizens, spread renewable energy information, and gradually create social learning and a greater awareness of energy issues (Kao 2016). In the end, one of the goals of the 'one person, one kilowatt' programme is that it will result in greater public participation in practical actions aimed at achieving energy democratization in Taiwan.

As of November 2015, HAND recruited 76 'Sun Shareholders' and secured 19 civilian roofs throughout Taiwan. Five roofs are in Northern Taiwan, six roofs are in Middle and Southern Taiwan and eight roofs are in Eastern Taiwan. HAND also organized a series of 'Sunday Sunday' information sharing meetings to explain more about the 'one person, one kilowatt' programme and encourage people to participate. In addition, many of the experiences and practices from Germany's energy transition are also introduced. Finally, HAND has also established several 'Little Architects' camps, which seek to provide children with knowledge about green building design, energy issues, and natural resources. Through these information sessions, as well as some hands-on implementation, the goal is to give young people some good information about renewable energy.

3.2 GRINNODOT

Huiping Chen is a sociologist who has investigated the ways that green energy technologies being utilized at the community level since she conducted her dissertation in Pingtung where community-based solar energy was utilized. Chen had been deeply affected by the 2011 Fukushima nuclear disaster, and saw how nuclear technology destroyed Fukushima. Yet, she believes that solar photovoltaic technology has great potential in Pingtung. Moreover, as a sociologist, she understands the effects that a disaster like the one that occurred in Japan can have on a community as well as the risks that are inherent with nuclear power. Therefore, Chen wanted to use her pioneering research interests to start a business, which led to a collaboration with two young men who are good at business and planning. Thus, in July 2015, GRINNODOT was formally established. In addition to Chen and her business experts, two engineers were also persuaded to join the team and take charge of cloud planning.

GRINNODOT has not only pioneered the idea of 'energy welfare', but also creates innovative business models as well. The company also encourages people to contribute funds for solar photovoltaic systems that will be used in buildings that house disadvantaged groups. Once installed, these systems generate income, through the government's FIT programme, which is used to support these same disadvantaged groups. In addition, although GRINNODOT is a business, the company upholds the ideals of a social enterprise. The company's main purpose is not for profit, but rather to create a platform, which they call SolarWe,[34] to network cooperative/individual donors and social welfare groups. It is estimated that through a simple donation for solar photovoltaic systems, in conjunction with the FIT policy, that the original amount given could show a return of 1.6 to 1.8 times.

The first project of the SolarWe programme was the 'SOS Children's Village of Taiwan'. This project aims to install 50 KW solar PVs on the rooftop of the building that houses the SOS Children's Village of Taiwan. Yet, in order to do this, they need to raise NT$3.5 million. Under the current FIT policy, it's estimated that NT$ 6 million can be raised within 20 years through the sale of electricity to the TP. In addition, an added benefit of this project is that it will eliminate 580,000 kg of GHG emissions, which equates to planting 4,100 trees. The project also trains children to be environmental narrators and explain the project to visitors. Undoubtedly, it's a unique form of environmental education for both children and adults.

Another platform, called EverDawn, operates like the 'one person, one kilowatt' programme. There are two types of participants in EverDawn: people who rent their roofs to GRINNODOT to install PV panels, and stakeholders in citizens' power plants who buy PV panels. After GRINNODOT determines where PV panels will be placed, they open a project platform on the EverDawn webpage. Then, they post all the information about this particular case, including the location, size/scale, expected monetary benefit, etc. EverDawn aims to make every roof become a power plant. In addition to seeking to activate idle rooftop assets, GRINNODOT also provides the opportunity for everyone to become a partner in citizens' solar power plants. People from all walks of life, including parents who wish to invest in their children's futures, office workers who are optimistic about green business opportunities, and even environmentalists who love nature may all join in this direct action to actively promote renewable energy. Unlike traditional models, people may buy one or more solar panels according to individual preference and then share in the green benefits with other participants in the project.[35]

3.3 Taiwan environmental protection union: the 'ten thousand sun temple' project

The Taiwan Environmental Protection Union (TEPU), one of leading environmental NGOs in Taiwan, has been an advocate against nuclear power since the

1980s. They also promote renewable energy development. TEPU even initiated the 'ten thousand sun temple' project in 2016. It's estimated that the amount of electricity consumed by temple lighting is over 21 million kilowatts a year. To put this in perspective, the total amount of consumption for electric lighting is 200 million kilowatts a year. In addition, the burning of incense at temples throughout Taiwan generates a great deal of carbon. Therefore, if each temple generates its own electricity, by installing solar PVs on temple roofs, it will not only save a great deal of money and electricity, but also provide local communities with electric power as well. Moreover, TEPU advocates that temples are often the best places to disseminate information and provide energy education in a community. In addition, TEPU also seeks to reduce carbon emissions by eliminating the burning of incense, which can help slow global warming. Lastly, TEPU encourages temples to use offerings to invest in community power plants, which will provide more benefits.

3.4 Free-roof project

Huei-Hsiang Hung, a sociologist and environmentalist, seeing the unjust and insufficient FIT mechanism, wants to produce electricity in the most efficient and environmentally friendly way. Hung generates solar energy and uses it directly, as selling solar energy back to TP's grid loses 20% of the original electricity produced.[36] Hung emphasizes that green energy is an alternative to coal-fired and nuclear energy, which brightens the state's energy future, but it's never cheap.[37]

Hung recruited 27 partners and raised one million NT dollars to buy solar PVs and install them on the roof of his Rainbow Restaurant.[38] He calls this 'free roof project',[39] in which he frees himself from TP and brings energy security in the long term. Solar electricity provides the power for six refrigerators and the air-conditioning system during the day, which helps Hung achieve a 70% self-sufficiency rate of energy. He estimates that the costs he incurred in installing this system will balance out within a decade, while achieving energy democracy.

Hung estimates that there are 8 million buildings in Taiwan, which means that there are many places to install solar panels. These panels can replace at least 40% of the electricity shortfall from retired nuclear plants. Moreover, instead of converting good farmland to set up solar panels in a large scale, Hung believes that it is better to develop small micro grids, such as in a community of 5,000 residents. The power generated from these grids may be used in-time and directly, which is in line with the original intention of developing green energy.[40]

Hung underscores the fact that electricity prices in Taiwan should be reasonably increased, not only to avoid wasting electricity, but also to increase incentives for green power generation. He also points out that there are two imperative measures that are needed to promote solar power generation on roofs. The first measure would be to relax financing interest rates for loans given for the purchase of solar panels, as current bank interest rates range from 3% to 5%. Hung suggests that the government subsidize 1% of the interest on these loans, so that the borrower will have less interest to pay. In this way, banks will not lose money and support will be given for green industry development. The second measure is that the government should start insurance coverage for solar PV businesses, in order to encourage more people to install PV panels on their roofs.

In addition to petitioning the government for change, the 'free roof project' is significant in terms of social advocacy. 'If the people don't want to live under the threat of nuclear risks and PM 2.5, everyone should do something about it.'[41] In the past, TP has had the final say about power policies, and the public has just passively accepted TP's decisions. Programmes like the 'free roof project' manifest the power of the people to actively participate in energy democracy and choose their own energy future.

V Strategic actions to accelerate energy transition: some suggestions

This paper seeks to uncover the reasons why energy transition in Taiwan has been sluggish and how civil society has advocated for energy democracy.[42] It is found that centralized, top-down approaches of energy transition are rooted in the legacy of long-term state-centred developmentalism in Taiwan. The transition to sustainable energy has been filled with vast uncertainties, in terms of technical opportunities and social risks. Unfortunately, the government, and TP's top-down system promoted by technocracy, tends to focus on solving short-term issues, and has thus failed to tackle the challenges inherent in the energy transition process. Currently, a number of public initiatives to build citizens' power plants have emerged. These endeavours are illustrated by social enterprises, such as 'one person, one kilowatt' and GRINNODOT, or through individual projects, such as the 'free roof project'. Yet, scholars need more time to observe and consider the outcomes and the prospects for these citizens' initiatives to promote energy democratization. However, in the process of achieving energy transition, it is imperative, without a doubt, to develop strategic actions that involve local governments, civil society, and business interests.

Energy transition for a city or community first requires the setting of targets that reflect a speedy energy transition effort. These targets include (1) reducing community-based greenhouse gas emissions for climate change mitigation (for example, reduce greenhouse gas emissions by 30% by 2035, compared to 2010 levels); (2) achieving energy efficiency through a reduction in energy consumption by 20% per person by 2035, compared to 2012 levels; (3) developing energy resilience by aiming to generate 10% of each local community's electricity through renewable sources by 2035.

Yet, in order to achieve these targets and become an energy sustainable city/community, several changes need to be made. We suggest that local governments make special efforts to encourage citizens and businesses to invest in energy efficiency, energy conservation, and clean energy solutions. To this end, a systematic, four-stage integrated approach would be helpful. Stage one involves education and outreach. In this stage, market transformation begins, as a significant segment of the population would develop a better understanding of things like renewable energy technology, green buildings, energy efficient vehicles, and thus place more value on them. However, until there is a threshold of demand, there is no market. As a starting point for the community programme proposed in the strategy, leaders need to ensure that market awareness exists, through education and outreach efforts.

Stage two concerns capacity building. One of the barriers to the widespread implementation of new and emerging energy technologies relates to the inability that most people have incorporating them into their lives. Ongoing learning is critical to the advancement of leading-edge, energy transition practices and technologies. In addition, local governments should assess the industry capacity that currently exists and help to facilitate improvements that may be required.

Stage three concerns incentive. In the early stages of a new market, uncertainty affects both suppliers and consumers. In these early-market situations, incentives should be provided to help reduce the risks and ease the anxieties that early adopters and suppliers may have. At the same time, financial incentive programmes should not become permanent programmes. The purpose of incentive programmes is to increase market penetration to a point where the product or service becomes the preferred option or a widely-supported option.

In Stage four, financial incentives are no longer required, as the market is sufficiently mature. Awareness, capacity, demand, and community support exist for the new energy initiatives. Sometimes, market forces alone will be sufficient to carry these initiatives. In other situations, regulations may be needed. While regulations would have been premature in earlier stages, they are appropriate in Stage 4, based on the new foundation that has been implemented.[43]

In addition, in both the upstream of formulating energy policies or the downstream to implement such policies, it is necessary to include varied subgroups, comprised of members of the general public, in order to develop a comprehensive energy transition policy and acquire procedural justice during implementation. This is particularly critical in dealing with conflicts between different values/camps of energy options in our society. Energy transition is not just a matter of technology or regulation, but also of social-psychological transformation. In other words, energy transition, on the micro level, relates to the values, environmental perceptions, knowledge, lifestyle, and family cultures of the people. This is why top-down institutional or regulatory changes alone seldom successfully achieve energy transition. Instead, change requires engagement with the public on several levels of government, while cultivating social learning and community empowerment.

We also learned that local community and NGO proposals eventually convinced the whole society that energy alternatives were both viable and possible. Therefore, it is important that each community/city creates and provides ongoing support for a community leadership body that will play a key role in leading local energy transition. It is suggested that a coalition of actors and effective multilevel governance will ensure the success of green cities. The most imperative fundamental enabling condition is a coalition of actors from national and local governments, civil society, the business sector, and universities, which are committed to advancing a green economy and its urban prerequisites, while making energy transition a top strategic priority. Another task for this coalition would be to promote long-term strategic planning as well.

Finally, we suggest that a citizen's panel be formed in order to deliberate on the challenges posed by local energy needs and climate threats. This citizen's panel could serve the following functions: (1) weigh the costs and benefits of each energy transition action; (2) employ public and transparent decision making processes so that citizens are confident that energy transition decisions serve the public good; (3) link city leadership to citizen education; (4) recognize and promote multiple reasons for energy transition.

To sum up, in the long run, the involvement of local communities, panel discussions by citizen groups, citizen fund-raising mechanisms, and social groups would serve as intermediaries to coordinate support for different projects that exist amidst various challenging social and political circumstances, which in turn may accelerate energy transition in Taiwan.

Notes

1 Please see: 'Nuclear Energy: Combating Climate Change', OECD Nuclear Energy Agency. https://www.oecd-nea.org/ndd/pubs/2015/7208-climate-change-2015.pdf
2 In the EU, governments are the initiator, while firms have to meet the goals established by regulators. In contrast, the US model allows firms to set their own standards, while government mainly plays a monitoring role. Both strategies basically make a trade-off between efficiency and welfare, which inevitably fall short in one dimension or another.
3 For instance, even the RWE project in Pingtung, which is considered to be a successful case, could hardly go further than the early stages because TP hesitated to upgrade their facility and redesign the grids. The original RWE was expected to expand through four stages; however, the second phase of RWE had yet to start after seven years, mostly because of TP's unwilling to further expand powerlines to rural areas. TP's passiveness is thus one of the greatest hindrances to the energy transition process.
4 Source: TP's webpage: http://www.taipower.com.tw/content/new_info/new_info-d16.aspx?LinkID=14. Retrieved on 4 June 2016.
5 The local TP offices held the final decision authority to approve applications for infrastructure placement, such as grid mergers.
6 While TP only charged $NT4 dollars per kilowatt, the FIT system requires that TP purchase renewable energy at about $NT 5–8 dollars per kilowatt, depending on its category.
7 The anti-nuclear DPP won the presidential elections in 2000 and 2016. However, the existing technocracy undermined former President Chen's 2004 proposal to discontinue the fourth nuclear plant. In 2016, the technocracy also convinced the new DPP government that Taiwan would not survive the energy transition without nuclear power (Ho 2006: 281–321).
8 Peer effects have proven to be quite influential in facilitating local renewables. Especially in an early market, peer effects and local organizations promoting PVs were identified as important explanatory factors for high local PV diffusion rates(Palm 2016).
9 Source: An interview with a former staff member from the Office of a Million Sunroofs, an office in ITIR. Apparently, the office has serious staffing problems. Although there are six branches in the office, there are only 12 staff members promoting their nationwide agenda.
10 The low ratio of domestic solar panels utilized in Taiwan also led to the perception that Taiwan PV companies were actually 'dumping' their solar panels by exporting them. Both US and EU courts in ruled in 2014 that they would collect anti-dumping taxes on solar panels exported by both Taiwan and China. Source: *Tianshia Magazine*, August 2014, http://www.cw.com.tw/article/article.action?id=5060156. Also, according to

an interview with the director of a renewable energy company, the domestic growth of renewable energy was only 8000 GW in the last decade.

11 Source: http://www.tri.org.tw/energy/Web%20information/2/2-1.pdf (in Chinese). Retrieved on 2 June 2016.

12 The Nationalist Party, better known as KMT, government eventually decided to terminate the controversial fourth nuclear power plant in April 2014 after strong opposition from civil society.

13 Source: http://shuchuan7.blogspot.tw/2011/03/blog-post_13.html (in Chinese). Retrieved on 10 March 2015.

14 Source: http://www.watchinese.com/article/2011/3074?page=show (in Chinese). Retrieved on 1 August 2016.

15 Source: Interview with environmentalist A on 22 October 2013.

16 In this section, the MOEA is the group that existed under the previous KMT administration (2008–May 2016).

17 Source: http://www.moeaboe.gov.tw/opengovinfo/govMain.aspx?PageId=govlist (in Chinese.) Retrieved on 15 December 2012.

18 Source: http://dailynews.sina.com/bg/tw/twpolitics/chinapress/20110322/06062315197.html. Retrieved on 11 January 2013.

19 Source: Interview with environmentalist L on 10 February 2013.

20 Source: http://www.moeaboe.gov.tw/opengovinfo/govMain.aspx?PageId=govlist (in Chinese). Retrieved on 21 May 2012.

21 Source: http://pnn.pts.org.tw/main/?p=24772 (in Chinese). Retrieved on 11 January 2013.

22 Source: Interview with environmentalist H on 29 February 2013.

23 Source: Interview with environmentalist H on 29 February 2013.

24 This statement was made by the Bureau of Energy, MOEA under the previous KMT administration (2008–May 2016). Source: http://www.sow.org.tw/natureNotesDetail.do;jsessionid=fc66178b4b53c0994575f8559fc0?id=52c4e8d0246fc543012479eace990014&notesSn=ff8080813630768b01369a6b684a0d23 (in Chinese). Retrieved on 28 April 2012.

25 Source: Interview with Environmentalist M in November 2013.

26 Source: http://e-info.org.tw/node/71417 (in Chinese). Retrieved in August 2013.

27 The implementation of the one-year programme (April 2015 to March 2016) varied in different areas, as the actual amount of subsidies depended on the actual rate of the energy-saving target.

28 These civil groups include: Green Citizens' Action Alliance (GCAA), Mom loves Taiwan, Citizens of the Earth, Taiwan, Taiwan Environmental Protection Union (TEPU), and Taiwan Renewable Energy Alliance (TRENA).

29 Source: http://e-info.org.tw/node/113020. Retrieved on 20 March 2016.

30 Source: http://www.appledaily.com.tw/realtimenews/article/new/20150714/647578/. Retrieved on 16 March 2016.

31 Source: http://www.csronereporting.com/news_2318. Retrieved on 20 March 2016.

32 Source: German Institute Taipei. http://www.taipei.diplo.de/Vertretung/taipei/en/08-Sustainability/Latest__News/Solarhaus.html. Retrieved on 15 March 2015.

33 In Taiwan, the annual electricity use per person is about 1000 watts. 'One Person, One Kilowatt' means that one generates electricity for one's own usage through this civil society movement towards energy democracy.

34 Source: http://www.solarwe.cc. Retrieved on 1 July 2016.

35 Source: EverDawn's webpage: http://www.everdawn.cc. Retrieved on 1 July 2016.

36 Hung chooses not to sell extra electricity back to TP's grid.

37 Source: Data from a focus group conducted by the authors on 14 October 2015.

38 The owner of Rainbow Restaurant buys organic products from local producers as a way to support local organic farmers. This restaurant is also designed to be a demonstration of solar energy to the public.

39 Source: https://www.newsmarket.com.tw/blog/72519/. Retrieved on 10 March 2016.
40 Source: Data from a focus group on 14 October 2015
41 Source: https://www.newsmarket.com.tw/blog/72519/. Retrieved on 10 March 2016.
42 Energy Democracy means ensuring that everyone has access to enough energy. However, the energy must be produced in a way that it neither harms nor endangers the environment or people. Concretely, this means leaving fossil fuels in the ground, socializing and democratizing the means of production, and changing our attitude towards energy consumption. Source: http://energie-demokratie.de/what-is-energy-democracy.
43 Source: https://www.edmonton.ca/city_government/documents/EnergyTransition Strategy.pdf Retrieved on 20 April 2016. More discussion on community-based energy transition, please see: http://energytransition.de/2015/03/danish-civic-participatio n-energiewende-lessons-for-germany/.

References

Adamsa, M., D. Wheelerb and G. Woolstona (2011). A participatory approach to sustainable energy strategy development in a carbon-intensive jurisdiction: The case of Nova Scotia. *Energy Policy*, 39(5), 2550–2559.

Chu, W. W. (1994). Import-substitution and export-led growth: A study of Taiwan's petrochemical industry. *World Development*, 22(5), 781–794.

Frantzeskaki, N., D. Loorbach and J. Meadowcroft (2012). Governing societal transitions to sustainability. *Sustainable Development, 15*(1/2), 19–36.

Geels, F. W. (2014). Regime resistance against low-carbon transitions: Introducing politics and power into the multi-level perspective. *Theory, Culture, and Society*, 31(5), 21–40.

Grundig, F. (2006). Patterns of international cooperation and explanatory power of relative gains: An analysis of cooperation on global climate change, ozone depletion, and international trade. *International Studies Quarterly*, 50(4), 781–801.

Hargreaves, T., S. Hielscher, G. Seyfang and A. Smith (2013). Grassroots innovations in community energy: The role of intermediaries in niche development. *Global Environmental Change*, 23(5), 868–880.

Hess, D. J. (2014). Sustainability transitions: A political coalition perspective. *Research Policy*, 43(2), 278–283.

Hess, D. J. and Q. D. Mai (2015). The convergence of economic development and energy-transition policies in state–government plans in the United States sustainability: Science. *Practice & Policy*, 11(1), 5–20.

Ho, J. C., C. T. P. Lee, S. F. Kao, R. Y. Chen, M. C. F. Ieong, H. L. Chang, W. H. Hsieh, C. C. Tzeng, C. F. Lu, S. L. Lin and P. W. Chang (2014). Perceived environmental and health risks of nuclear energy in Taiwan after Fukushima nuclear disaster. *Environment International*, 73, 295–303.

Ho, J. C., S. F. Kao, J. D. Wang, C. T. Su, C. T. P, Lee, R. Y.Chen, H. L.Chang, M. C. F. Ieong and P. W. Chang (2013). Risk perception, trust, and factors related to a planned new nuclear power plant in Taiwan after the 2011 Fukushima disaster. *Journal of Radiological Protection, 33*(4), 773–789.

Ho, M. S. (2006). *Why the DPP government's nuclear phase-out policy failed? Green Democracy: A Study of the Environmental Movement in Taiwan*. Taipei, Group learning publisher. (in Chinese).

Johnson, C. (1987). Political institutions and political performance. In F. Deyo. *The political economy of new Asian industrialism*. Ithaca, Cornell University Press.

Kao, S. F. (2016). Local practice of energy transition: Community-based energy and people's power plant. In Kuei-Tien Chou and Tze-Luen Lin (Eds.) *Fourteen Lectures of Energy Transition in Taiwan* (pp. 181–197). Taipei: Chuliu Publisher. (in Chinese).

Kemp, R. and D. Loorbach (2006). Transition management: A reflexive governance approach. In J. P. Voß, Dierk Bauknecht and René Kemp, *Reflexive governance for sustainable development*. Cheltenham: Edward Elgar.

Khan, N., E. Asproudis and N. Kakabadse (2014). The transatlantic regulatory timing of policy innovations: The case of renewable energy. In S. Schepers and A. Kakabadse. *Rethinking the future of Europe – A challenge of governance* (pp. 118–138). New York, Palgrave Macmillan.

Lee, M. (2011). Reasonable and practical approach: Nuclear energy – a choice we cannot give up. *Science Monthly*, 494, 108. (in Chinese).

Liang, S. W. and Lee, C. Y (2014). Risk perceptions and attitudes of Taiwan citizens regarding nuclear power generation based on NIMBY and cognitive dissonance theory. *Taiwan Journal of Public Health*, 33(4), 428–444. (in Chinese).

Lund, H. (2014). *Renewable energy systems, second edition: A smart energy systems approach to the choice and modeling of 100% renewable solutions*. Oxford, UK, Academic Press.

Palm, A. (2016). Local factors driving the diffusion of solar photovoltaics in Sweden: A case study of five municipalities in an early market. *Energy Research and Social Science*, 14, 1–12.

Pralle, S. and J. Boscarino (2011). Framing trade-offs: The politics of nuclear power and wind energy in the age of global climate change. *Review of Policy Research*, 28(4), 323–346.

Rifkin, J. (2013). *Third industrial revolution: How lateral power is transforming energy, the economy, and the world*. New York, Palgrave Macmillan.

Schweizer, P.-J., O. Renn, W. Köck, J. Bovet, C. Benighaus, O. Scheel and R. Schröter (2014). Public participation for infrastructure planning in the context of the German 'Energiewende'. *Utilities Policy*, 1–4.

Seyfang, G., J. J. Park and A. Smith (2013). A Thousand Flowers Blooming? An examination of community energy in the UK. *Energy Policy*, 61, 977–989.

Szulecki, K., A. Ancygier and D. Szwed (2015). Energy democratization? Societal aspects of de-carbonization in the German and Polish energy sectors. ESPRi Working Paper, 5.

Wade, R. (2003). *Governing the market: Economic theory and the role of government in East Asian industrialization*. Princeton, Princeton University Press.

Yang, C. Y. (2013). A cross-straits comparative study of low-carbon economies. *Hwa Kang Journal of Agriculture*, 31, 27–45. (in Chinese).

Yang, W. S. (2011, October 31). Hope to decrease nuclear power gradually. How does Ma respond? *United Daily News*. (in Chinese)

6 The grassroots movement for energy transition in Korean society

Jin-Hee Park

Introduction

Korea ranks 7th for greenhouse gas emissions, and electricity produced from nuclear power plants is 4th following America, France, and Russia. Nuclear energy accounts for 30%, bituminous coal 38%, LNG 22%, and renewable energy 2.8% of energy production in 2014, representing an energy system based on fossil fuels. In primary energy, oil is 37%, bituminous coal 27.8%, nuclear energy 11.7%, and combined waterpower and new renewable energy 4.5% (KEEI, 2015). Electricity produced mainly from fossil resources causes a lot of social problems. The more greenhouse gas emissions increase, the more responsibility for climate changes grows. Atmospheric pollution caused by coal use increases, there is a high risk of nuclear accidents, and there are social conflicts related to the construction of nuclear waste disposal facilities.

According to the Paris Agreement, Korea made a plan to reduce greenhouse gases and stressed the transformation of a fossil-fuel oriented energy system. In the Second national energy plan, the enforcement of energy demand management and the extension of distributed power plant systems are emphasized. The government, however, has pursued the plan to construct 11 nuclear power plants by 2024, which will make the generation of nuclear electricity equivalent to 29% of total electricity production. Although, among OECD countries, Korea has the least renewable supply rate, Korea sticks with its plan to increase its renewable energy supply rate to just 11% by 2035. After the Fukushima nuclear accident, though there has been public concern about additional construction in the nuclear power plant concentration area, the government continues to push ahead with its policy. Construction of electricity transmission lines is essential to send electricity from the nuclear power plants concentrated in the southeast to the capital area which demands high electricity consumption. Due to this policy devoted to nuclear power, in 2013 Miryang residents' fight against the construction of high voltage transmission cable towers became intense. The energy policy of government giving priority to economic feasibility promotes coal power plant development whose electricity production cost is second lowest after that of nuclear power plants. As a result, this has a serious impact on health, increasing particulate matter concentration, which means Korea has to pay a greater social cost because of an energy system centred on fossil fuel.

There are voices urging Korea to fundamentally reflect on the present energy system and to transform the system. Since anti-nuke movements in the 1980s, Korean citizens have required an energy transition. Citizens who had participated in the anti-movement against the construction of nuclear waste disposal facilities warned about the risk of a nuclear accident and requested the closing down of nuclear power plants. However, this anti-nuke movement, which originated from the environmental movement, failed to evolve into a citizen movement and prevent the enhanced nuclear energy policy of the government. In 2000, Korean citizens began to participate in an energy transition movement which evolved into a post-nuclear and local energy transition movement after the Fukushima accident. This paper studies how the movement developed, how energy transition movement is going, who initiated energy transition movements, and what the task of the energy transition movement is, in particular, dealing with the development process of the energy transition movement related to the formation and growth of energy citizenship in Korea.

1 Energy citizenship

Citizenship, which is the status of a person recognized under the customs and law including their right to freedom, politics, and society, has developed into the concept embracing modern society issues such as a republican perspective, and environment and ecosystem problems, which represents a public right, duty, virtue, and responsibility (Leach and Scoones 2005; Hong and Lee 2014). Marshall who theorized the concept of citizenship insisted that social inequality would improve through development of citizenship and, that civil (including rights to freedom), political, and social rights would be required (Marshall 2009). He placed a premium on the rights to freedom giving us a choice to choose religion or ideology, civil rights giving us an opportunity to participate in a free market economy and protect ourselves, and political rights giving us the opportunity to participate in political life. Above all, he considered social rights as the most important. The concept of Marshall's citizenship changes comprehensibly in modern society where environment issues and scientific-technical abuse arise. And environmental, ecological, and science and technological citizenship are parts of its changing. That is, the concept of modern citizenship includes all these ideals. Before details of the energy citizenship concept are considered, an explanation of environmental, ecological, and science and technological citizenship are categorized as follows.

Bryan Turner, Newby, Van Steenbergen and others criticized Marshall's citizenship ideals which considered the concept of rights as important, arguing for a new citizenship arising out of the property of modern society deriving from changes in the relationship between humans and the environment. They insist that there should be a fourth dimension that adds civil rights and responsibilities concerned with environment and sustainability (Flyn *et al.* 2008). While Marshall argued that rights to freedom, politics, and society are fundamental, they insisted environmental citizenship acting as a fourth dimension be included in this existing definition of citizenship. In the concept of environmental citizenship, the right of the citizen is expanded to the right that can request the government to protect people from environmental pollution.

Based on this concept, Andrew Dobson developed the present environmental citizenship emphasizing environmental responsibility. According to Dobson, every citizen has a responsibility to protect environment such as air, water, and nature, preserve biodiversity, and restrain environment abuse to maintain a healthy environment (Dobson 2007). And environmental citizenship emphasizes the importance of responsibility for the community over individuals, which is close to republican citizenship. Citizens fulfilling environmental citizenship share the responsibility to protect nature and recycle to realize a sustainable society.

Although the new concept of 'environmental citizenship' made a contribution to meaning expansion of citizenship, it became clear that the environmental crisis in modern society could not be sufficiently formulated with this new concept. 'Environmental citizenship', being based on the existing citizenship of which rights and duties are connected with mutual contracts, did not solve the problems of future generations and other species threatened from an environmental crisis. In other words, 'environmental citizenship' grounded on the concept of citizenship that citizens do their duties due to being devolved rights from the nation lacks responsibility for future generations and other species which are not guaranteed any rights. That is why Dobson introduced anew concept called 'ecological citizenship' (Dobson 1998). Ecological citizens realize that ecological citizenship has a responsibility to protect future generations and other species, but will not be guaranteed any rights from them. In the concept of ecological citizenship, virtue and compassion of citizens are also considered as important without classifying a public or personal area. Responsibility and implementation to protect future generations are recognized as equally significant regardless of home or political community. It means that, if ordinary consumers are involved in environmental protection activities, they put their ecological citizenship toward a sustainable society into practice. Realizing ecological citizenship can be reinforced through government policy or education (Melo-Escrihuela 2008).

In modern society, ecological and environmental crisis cause the concept of citizenship to evolve. Likewise, the relation between science technology and society brought some changes to the rights and responsibility of citizenship. Science and technology cause risks and the knowledge of experts becomes less reliable. That is because citizen participation is essential for 'risk society', whose risks cannot be controlled anymore by experts on their own. Uncertainty of knowledge is shown as evidence of citizen participation in policy decision-making, eliminating the hierarchical order between expert and lay knowledge. As new rights of citizenship evolve, assurance on the rights of citizen participation, and the institutionalization of citizen participation are discussed in various ways. Through this process, the concept of 'science-technology citizenship' is newly constructed by the practising citizen jury, consensus conference, or citizen science. 'Science-technology citizenship' can be defined as citizenship aiming for democratization of science-technology, which means that a citizen has rights to access freely science-technology information and to participate in policy decision-making and risk management in the context of risk society (Lee 2011).

As Citizenship has an effect on energy issues related to climate change, a new concept named 'energy citizenship' has emerged. British environment-sociologist Devine-Wright first used the term 'energy citizenship' during his research on voluntary behaviour and attitude required by citizens for establishing a sustainable low carbon energy system. He explained that the policy maker or engineers who depend on the existing centralized energy system regarded the public as deficient citizens who do not care and have no knowledge about energy (Devine-Wright and Devine-Wright 2005). He also accounted for the reason that 'energy citizenship' appeared because citizens as interested parties began to be aware of energy consumption and claim their rights. Developing decentralized energy systems such as small wind power generators or block heating systems fuelled by biomass, can better help citizens understand energy production, supply, and consumption, and recognize climate change resulting from energy consumption. Also, if the government provides participation opportunities for citizens as advisers of regional renewable energy projects, they can actively take part in energy policy decision-making. 'Energy citizen' represents people feeling responsible for climate change, emphasizing equality, justice in relation to issues of energy poverty or conflicts about power plant site, and fulfilling 'energy citizenship' that promotes joint renewable-energy projects as energy cooperatives (Devine-Wright, 2007: 72–80).

Similar to appearance of science-technology citizenship involving policy and technical development, energy citizenship with ownership of the facilities also tries to engage in energy production and sustainable energy transition. Like environmental and ecological citizenship, energy citizenship put realizing a sustainable society before exercising rights. In fact, energy citizenship has evolved from saving energy in a personal capacity to being equipped with community energy facilities and taking part in energy policy decision. Devine-Wright said that renewable energy development helps citizenship mature and, according to the energy policy of the British government, enhances the new form of citizenship (Devine-Wright 2007: 74). Energy movements in Korea, in particular the grassroots energy transition movement can be read as a formation of energy citizenship.

2 The emergence of the energy movement

Korea has been aware of energy issues socially since the late 1980s when children with anencephalus related to nuclear power plants were born and social conflict deepened over nuclear waste disposal sites. In Yeonggwang-gun,[1] there were a series of miscarriages of foetuses with anencephaly and children born with macrocephaly, which led to a damage compensation movement. The residents established a committee for compensation with the assistance of some experts and required physical examination, emigration measure, and damage compensation (*Dongailbo*, 10 August 1989). As Korea has depended on nuclear power which reached 46% of total electricity production since late 1988, the problem of disposal of nuclear waste has become an issue. Accordingly, the Ministry of Science and Technology and Ministry of Energy and Resources chose three proposed sites for nuclear waste disposal facilities and geologized. This secretive site selection aroused Yeonggwang-gun residents' fierce opposition. Eventually, the government stopped geologizing and rejected the site selection (Hong 2016: 193). This kind of site selection happened again in Anmyeon-Island, resulting in the residents' 'resistance of seven days' against it, and the government withdrew its plan. In addition, the residents' violent demonstration made the headlines in national daily newspapers, and the issue of nuclear waste disposal facilities was taken seriously. The National Anti-Nuclear Network organized by environmental movement organizations and local community groups made a request to the government to provide transparent construction plans for nuclear waste disposal facilities and, at the same time, make a plan for eventual post-nuclear energy policies (*Hankyoreh*, 6 November 1991).

Despite those anti-movements of local citizens and environmental organizations, the Korean government unilaterally pushed Gureop-Island nuclear waste disposal facilities once again, which led to an opposition movement against its plan. Not only Gureop-Island residents but residents living in Incheon took part in the movement. However, unlike the cases of Yeonggwang-gun and Anmyeon-Island, the government did not withdraw its plan at first. Through administrative organization, the government publicized its regional development plan systematically and promised welfare policy to try to prevent the opposition demonstration. The government attempted to induce the residents not to demonstrate. The conciliatory policy seemed to have an effect, but the selected site was no good as an active fault was found at the site of the proposed facilities (Hong 2016: 204). The promotion and economic support policies which the government tried have developed since then. To sum up, the government chose the method to select the site of nuclear waste disposal facilities through providing an economic incentive. In spite of modifications in the nuclear energy policy, in 2003, large-scale demonstrations were held against nuclear waste disposal facilities in Buan-gun, frustrating this site's selection by the government.

The movement against nuclear waste disposal facility construction became a trigger to recognize the dangers of nuclear development, but does not help to solve the problems. Though this movement delayed the policy and the construction, the nuclear waste disposal facility was eventually constructed in Gyeongju-city in 2005. What's worse, it was not issued in the public domain that the planned construction of nuclear power plants should be stopped in order to prevent social conflicts. The Korean Federation for Environmental Movements was established in April 1993 under the slogan 'Transition toward environment-friendly and renewable energy system', which was however not a core activity of the federation (Korean Federation for Environmental Movements 1993).

It was in 2000 that problems of existing energy supply system were focused on by NGOs. The Center for Alternative Energy organized to deal primarily with the alternative energy issue was separated from Korean Federation for Environmental Movements, the large national environmental NGO. The Center for Alternative Energy (CAE) argued that the energy crisis could be solved not by the intensive use of nuclear energy and large fossil-fuel reserves but by the countermeasures of effective energy saving and active renewable energy development. The CAE advocates establishing a decentralized and sustainable energy system and initiated the movement (CEA 2000). The CAE began to develop technical systems such as a civil photovoltaic power plant built with investment by the members. In May 2003, the first civil power plant having a power capacity of 3.06kW was built. The CAE tried also to found an incorporated company to promote decentralized photovoltaic power plant construction. The civil power plant construction movement had an effect on the activities in Buan-gun concerned with an alternative to the nuclear power plant, leading to the building of photovoltaic power plants through civil investment in Buan-gun.

The civil plant building movement, at the instigation of Center for Alternative Energy, could gain popularity by introducing and institutionalizing Feed in Tariffs (FIT), and by reorganizing the related policies. With renewable energy oriented policy being emphasized internationally, the Korean government, in 2002, introduced Feed in Tariffs designed to allow electricity produced from renewable energy sources to be purchased at a fixed price (Koo 2013: 3–4). This system ensures that electricity generated from renewable power plants can be sold at a relatively high price, creating a favourable environment for civil power plants using renewable energy. It turns out however, the system is not the best solution for the small civil power plant, since, under the law related to electricity transmission, civil plant providers should register at Korea Power Exchange at that time to sell its electricity just as large providers do. That means, to sell electricity, the plants should be connected to the power system with relative high cost, which is economically unfavourable for small plant providers. Therefore, the CAE called for amendments of the relevant laws and actively encouraged the expansion of civil power plant movement (Chang and Han 2008: 102–103).

In October 2005, besides the member plants of the Center for Alternative Energy, three civil solar power plants were built in Buan-gun, which was seeking alternative energy after the movement against nuclear waste disposal facility. The residents in Buan-gun and environmental organizations constructed civil power plants resulting from 'achievements of the movement against nuclear waste disposal facility for 2 years' (Green Korea 2009). In 2006, Korea YMCA established a 200kW civil solar power plant in Suncheon which was large scale at that time, and then Daegu civil organizations and Ulsan Federation for Environmental Movements continued the establishment of civil solar plants in 2007 and 2008. The Center for Alternative Energy, which started to build civil plants, now renamed the Center for Energy Transition attempted to expand the number of solar power plants through members' investment. Since then, by 2015, the Center has constructed 40kW solar power plants to help the residents generate electricity in Goesan and Cheongyang-gun. Civil solar power plant movements triggered by the activity of the Center for Alternative Energy was the first movement to enable regional residents to prepare a countermeasure to nuclear development and climate changes. The construction of civil solar power plants gave opportunities to citizens to take part in energy production.

3 Discourse of 'energy transition' and attempts of local energy transition

In parallel with the establishment of civil plants, some villages took the initiative by themselves to solve their local energy problems. With the assistance of the residents who led the opposition movement of Buan nuclear waste disposal facilities, a 3kW civil photovoltaic power plant was built in Deungnyong village in 2005. Beyond opposition to the nuclear power plant, the energy transition movement aiming to produce alternative energy by citizen themselves arose at village level. By 2015, Deungnyong village, after building renewable energy plants, set a goal that 50% of total energy consumed in the village should be produced by using sunlight, wind power, or biomass, and conducted energy transition experiments. Since 2008, Jeonbuk Agenda 21, Green Korea, and Korea Energy Agency have participated in the energy transition movement of Deungnyong village utilizing energy conservation, house insulation improvement services, civil solar power plants, and geothermal power plant establishment (Lee 2014: 129). This first 'energy transition' attempt at village level in Deungnyong received an increasing amount of media attention and inspired the concept of 'energy self-sufficient village' in Korea. And, this experiment of energy transition was enabled thanks to the activists in local communities, various civil groups, and government cooperation.

Meanwhile, Feed in Tariffs introduced in 2002 offered an opportunity to the residents to participate as energy suppliers. Feed in Tariffs can be understood as an opportunity which produces corporate profit by investing a village common fund in solar power plant. In 2008, Jeju-do, Hwasun-ri registered itself in corporation and established 'Beonae Solar Power Co' by investing the Hwasun-ri common fund. A total of KRW 1,600 million common fund, which was damage compensation received due to heated effluent of the thermal power plant was invested and can be returned in 10 years thanks to Feed in Tariffs. As the photovoltaic power plant project began to utilize the village common fund, it resulted in a rise of the contributing residents' recognition about energy production as they experienced electricity production at a village level. Similarly, residents in Gangwon-do Inje-gun Namjeon 1 ri constructed a photovoltaic power plant to increase the profits of the village common fund. The head of village and other leaders who took a class of leadership training founded 'Namjeon 1 ri Farming Cooperative' with a bank loan and the village development fund, and built a 300kW photovoltaic generation facility. The profit from selling electricity was disbursed in the village joint businesses, increasing the welfare of the village (Lee 2014: 133–134).

The energy transition of Deungnyong village and the photovoltaic power plants projects of the two villages were accomplished by the village leaders and NGOs. In Korea, energy transition attempts were also conducted under private partnership, one of which was Yeondae-Island eco-island development project pushed ahead by Green-Tongyeong 21 and Tongyeong municipal government in 2007. Green-Tongyeong 21 (Local Agenda Group) demanded that Yeondae-Island should be declared an ecological village and reformed. To realize this project suggested by a

Local Agenda organization, Tongyeong municipal government enacted a 'Local Law for Fostering Eco-Island', encouraging the residents committee to take part in the eco-island project, and secure a budget for photovoltaic power plants needed for the ecological village. Sixteen members of the committee including six residents, government employees, Green Tongyeong 21, local media, and experts set up the action plan; the plant site was selected by the residents. Through this process, fifty 3kW photovoltaic power generators were installed including an eco-experience centre equipped with a solar panel roof and bicycle dynamos. Green-Tongyeong 21 provided education and experience opportunity to learn renewable energy facilities, and allowed the residents to be involved in decision-making and project management (Kang and Lee 2016: 159–162). In this way, the eco-island project was conducted practically by citizens' participation. Though it was a typical top-down project at the start, the cooperation between Tongyeong municipal government and Green-Tongyeong 21 allowed the democratic implementation of this energy transition experiment. Consequently, residents who received the energy and/or plant management training paid more attention to photovoltaic power plant or electricity consumption, raising awareness of energy issues.

Thanks to those local energy movements, in the late of 2000s, the civil plant building movement has been transformed into an 'energy self-sufficient' or 'energy transition movement'. Like Buan, in some locales, energy transition was conducted through the cooperation of local activists, NGOs, and government institutions, or the investment decision on the installation of a photovoltaic power plant by leaders of the village. The case of Deungnyong village was very significant because it was the first attempt that a village became the main agent of energy production by setting up energy plans at village level and by installing renewable energy facilities. With regard to evaluating the residents' role in the process of the attempts, it was far from a grassroots movement for energy transition. The residents were just beneficiaries who involved themselves in the energy conservation campaign that Green Korea led and benefited from the government's house insulation project. The installation of photovoltaic panels and geothermal heat equipment in the village was not the result of direct investment by the residents. Unlike the case of community energy in Great Britain, the strengthening of awareness based on ownership of energy production facilities seemed to have failed in Korea (Lee and Han 2014: 84–85).

In the cases of Namjeon 1 ri and Hwasun-ri, energy transition of the village was not its ultimate goal, however, with the residents' participation in investment decisions, the residents could make a contribution to the technical change in the village's energy system by producing solar electricity. Taking part in building a photovoltaic power plant meant the residents raised their understanding about electric power and attention to local energy issues. As Feed in Tariffs guaranteed a high profit for operation of the photovoltaic power plant, the residents' attention on photovoltaic power could be sustained. That is, FIT has played an essential role in those local energy transition experiments. However, the experiments of two villages could not be developed further beyond building a photovoltaic power plant, because there was no plan for energy transition in the village. In the case of

Yeondaedo unlike Deungnyong village, community participation was strengthened and, at the same time, provided opportunities to experience energy transition. The residents were involved in site selection and were aware of the significance of energy transition. However, the residents' participation in facility investment was restrictive, which made it hard to keep their attention. As the support of NGO and/or local governments decreased, the energy transition attempts didn't last. The attempts that were not from the local residents' intrinsic motivations face a situation in which the attempts will have to be stopped if the support is cut.

4 The formation of energy citizens

The Fukushima nuclear accident that happened in March 2011 brought qualitative change into the grassroots movement for energy transition in Korea. The nuclear accident made Korean people frightened of nuclear fallout and paved the way for deep reflection on nuclear energy policy. Instead of Anti-nuke, the concept of Post-nuke appeared in the discourse about nuclear energy policy. Breaking from the movement against nuclear waste disposal facilities or nuclear power plants with no countermeasures, a rational solution of phasing out reliance on nuclear energy needed to be found. After the accident, apart from the existing environmental movement organizations, new forms of civil organizations were founded by housewives like Child Save. They began to monitor radioactive pollution of food by themselves. Not only did regional citizens' alliances form post-nuke countermeasure committees, but professional groups were organized by professors and/or lawyers under the slogan of 'Post-nuke'. The new organizations claimed to advocate energy transition that allows renewable energy uses to expand and, ultimately, construct an energy system freed from nuclear energy (H. S. Lee 2016). With civil activists and some experts, the scenario about alternative energy including a phase out of nuclear power plants was developed. With the help of civil activities like this, the opposition party and progressive parties of the national assembly began to prepare a new post nuke energy policy, while some municipal governments announced that they would implement a post-nuke energy policy. Post-Nuke movements served as the foundation for deployment of the grassroots movement for energy transition. After the Fukushima accident, citizens involved in the existing organizations or new groups actively initiated electricity saving movements and created solar power cooperatives to set up a new energy system. These grassroots movements can be classified as the movement of renewable cooperative building and the community movement for the energy self-sufficient village.

1 The foundation of renewable energy cooperatives for energy transition

After the Fukushima accident, citizens who were engaged in ecological environment movements took the lead in building renewable power plants by founding cooperatives. Above all, the members of Hansalim Cooperative, Won Buddhism Environment Network, and Korean Federation for Environmental Movement in

Seoul were active. These organizations have sought practical countermeasures for Post-nuke like radioactive contamination monitoring. What they really paid attention to was the Cooperative Act announced in 2012. The law which made it easier to form cooperatives with at least five members let small investors take part in the installation of photovoltaic power generation. The members interested in Post-nuke were inducted as cooperative initiators and founded solar power cooperatives. The cooperatives of photovoltaic power plant established in late 2012 constructed photovoltaic power plants that range from a scale of 19kW to 500kW on the roofs of borrowed sites, schools, or public buildings. Through these attempts, citizens could engage in the process of energy production both through cooperatives as well as civil power plants. This means that citizens got an opportunity to join in energy transition through direct participation (Park 2015: 197–200).

By 2013, installation of photovoltaic power generation as energy transition was fulfilled with the assistance of civil organizations, small-sized local groups, and ordinary citizens. The Sun and Wind Energy Cooperative was founded in Seoul, Eunpyeong-gu by citizen activists including Nowon Sunlight and Wind Cooperative in Seoul, Nowon-gu. Especially, in the case of Nowon Sunlight and Wind Cooperative, with the active participation of the head of Nowon local government district, many Nowon-gu officers joined the cooperative, making this the largest cooperative in Seoul. Except for these cooperatives, citizens involved in environmental movements in Su-Won organized a social cooperative to donate the profit from solar power sale to welfare services. This combines public interests from Post-nuke and social welfare. Around the same time, photovoltaic power plant cooperatives in Anyang, Gunpo and Uiwang-si were also founded. Besides the capital area, some cooperatives were set up in Gyeongnam, Ansan, and Jeonbuk. The number of citizens involved in the seven cooperatives in Seoul reached 2,306 as of August 2014. In particular, the national cooperative, Hansalim has 1,309 members engaged in building solar power plants. The plant capacity installed by cooperatives was 333kW in the case of the seven cooperatives in Seoul, and Hansalim represented 531kW (Park 2015: 191). Like the civil power plant movement, the foundation of solar power cooperatives was another citizen action for Post-nuke society. However, energy production participation through cooperatives was not easy in Korea at that time. After the Fukushima accident, photovoltaic power plants as an alternative to nuclear energy drew the attention of citizens who recognized the risk of nuclear energy. As a result, fundraising for the first cooperative-built solar power plant was not as difficult as might have been expected, while most cooperatives found it difficult to secure building sites, except Hansalim for which a warehouse was used as the power plant site. They needed to obtain the principal's approval to use a roof of a school as the power plant site. However, a principal who has a low opinion of photovoltaic power plants often rejected cooperatives' construction plans. Though the Seoul metropolitan government supported such initiatives, for instance providing information about the sites of public buildings, the projects were frequently abandoned during the contract process with building owners. Even if they borrowed the site, because of the high grid connection cost, there were some instances where construction was impossible. In addition, under Renewable Portfolio Standard (RPS) which has replaced FIT since 2012, Renewable Energy Certification (REC) was too low priced to make up profit of cooperatives, which led sluggish sales. As a result, the cooperatives hardly received their expected profits from photovoltaic power plants (Park 2015: Yun and Shim 2015). The reasons why European countries like Germany could expand their renewable energy cooperatives are incentive policies like an exemption from the grid connection cost for small solar power plants or easing of costs. They also have FIT instead of RPS. Unlike these countries, the support system for renewable energy in Korea makes growth hard for cooperatives. After most cooperatives in Korea were organized, they just constructed a first photovoltaic power plant and did not make any further progress in development, which did not lead to profits. Consequently, the foundation of solar power cooperatives has plateaued since 2015.

The decline of cooperative activities has resulted in the constrained formation of 'energy citizen' through cooperative organization. The members of cooperatives could grasp better the relationship between climate change and renewable energy, and raise their awareness of energy consumption. However, the low activity of cooperatives in building solar power plants prevents the members from understanding the energy issues. Daily activities or education offered by cooperatives are often suspended, because the projects of cooperatives could not progress. It is reality of cooperatives in Korea that, except for founders who have a clear vision of energy transition, the rest of the members did not develop into true energy citizens. Though the emergence of the founders could be considered as an achievement for the energy movement, there is no doubt that the development of the energy transition movement depends on the growth of the 'energy citizen'.

2 Community movement for the energy self-sufficient village

With the foundation of cooperatives at the level of dong or apartment blocks, citizens sharing the area began to take part in the energy movement. Unlike Deungnyong village or Yeondaedo eco-island project, a community movement for energy self-sufficient villages also started in Seoul and capital areas after the Fukushima accident. In January 2012, 'Sungdaegol People' living in Dongjak-gu initiated the Negawatt movement which expanded to an energy transition movement. The residents involved in the children's library building, after the Fukushima accident, invited a person from an NGO to give lectures about energy. Those activities enabled a formation of Negawatt movement. Gathering the residents voluntarily involved in electricity saving at home, they organized a small group of households that present the amount they save monthly in the village library. This action was named the Negawatt Movement. This means that electricity saving could lead energy production without constructing any other power plant (Y. J. Lee 2016). The media helped spread the message of this movement to other communities.

The Negawatt movement of the Sungdaegol people evolved into a community movement for an Energy Self-Sufficient Village in September 2012 based on the new policy of the Seoul metropolitan government. In April 2012, Seoul formulated the policy 'One less nuclear power plant' and implemented it for the first time among local governments. The policy aimed to settle the 'environmental injustice' problem by reducing nuclear power plants one by one, and by increasing self-sufficiency in Seoul. Detailed policies involve a home energy clinic service and supporting community movement of energy self-sufficient village.[2] Sungdaegol enforced this home energy clinic service on 2,000 households in Dongjak-gu by utilizing this policy actively, and won a public contest, 'Community movement for Energy Self-Sufficient Village' of Seoul. They carried out the insulation of buildings with the prize money. The electricity saving movement for households extended to local stores and schools, and, at the same time, engaged in energy production. With assistance of Alternative Technology NGOs, 'Sungdaegol people' organized the installation of solar thermal fan heaters and a wood stove 'community school' building. In addition, energy classes for the future generation were opened for local middle school students, and the village energy festival was held to raise residents' recognition of energy.

Achievements of 'Sungdaegol people' are the establishment of 'Community. Oikos cooperative' in 2013 and Sungdaegol Solar power cooperative. 'Oikos cooperative' was formed by local residents like the carpenter, architect, and boiler engineer to establish the insulation of buildings as a community business. The cooperative carried out insulation improvements at the senior community centre and four detached houses, and opened an 'energy super market' which is selling high energy efficiency products and providing a place for residents to increase their awareness of energy. The people concerned with Oikos cooperative, in 2014, became involved in a Housing Environment Improvement Project supervised by Seoul Metropolitan Government, and had energy issues contained on the project. With the aim of constructing a photovoltaic power plant, Sungdaegol founded 'Sungdaegol Solar Power cooperative', and carried forward energy production (Y. J. Lee 2016).

Sungdaegol's Energy Self-Sufficient campaign shows that an energy movement limited to a field such as electricity production developed into a house insulation service and thermal energy issues, which means an actual energy movement was launched by citizens. As the activities of Sungdaegol people, unlike previous civil energy movements, were on the basis of residents' spontaneous participation and extended to power saving, energy production as well as energy education, its movement can be regarded as significant progress. Energy education provided by Sungdaegol people contributed to 'Community movement for Energy Self-Sufficient Village' spreading to other villages, with not only the residents but other local government officials joining in the class. Activities in Sungdaegol proved the emergence of an energy citizen who realizes energy transition.

There is another movement called Kwanak-gu eco-village similar to Sungdaegol people. After the Fukushima accident, Kwanak Resident Network located in Seoul Kwanak-gu recognized the importance of resident activity to change the energy production structure into renewable energy and reduce energy excessive consumption. In 2003, the network founded a group of residents named 'Dream Village Eco Wind', who are responsible for projecting energy transition at community level, and also organized the 'Kwanak Energy Environment Network' to fulfil its project. In Kwanak-gu, eight Negawatt plants were built to encourage the residents to save energy, and regular 'Energy education and Campaign' was offered to improve recognition of energy environment. In 2013, there were 22 opportunities for the residents to receive energy education. The participants in this education programme had signed up for eco-mileage Seoul which was initiated for the promotion of energy savings. In 2014, energy clinic services were conducted on 80 households. In 2015, installation of a mini-sized solar panel in apartments was pushed ahead, which resulted in installing solar panels on eleven households. And a promotion committee for the photovoltaic power plant cooperative was created as well. Like Sungdaegol, Kwanak-gu eco-village supported by Seoul city could launch an Energy Self-Sufficient Village project. These activities achieved 9% of energy saving and more than 400 households joined eco-mileage to save energy. And an energy super market as well as a periodic energy festival was also in operation (Cho and Yun 2016). A civil energy movement led by local residents resulted in power-saving and energy production by installing mini-sized photovoltaic panels and the foundation of photovoltaic power plants cooperatives.

Energy Self-Sufficient Village supports the measures of Seoul to promote the residents' groups of apartment complexes to come forward with electricity saving movements and solar power generation by mini-sized solar panels. In 2013, Hyundai Prumi which is a resident organization of Sindaebang Hyundai Apts in Dongjak-gu put energy saving into action, influenced by Sungdaegol activity. After selection as an energy self-sufficient village in 2014, the organization has run an apartment energy clinic service, energy super market, installation of the roof photovoltaic power plant, and an energy experience centre. Seongbuk-gu Seokgwan-dong Doosan Apts, Jongno-gu Changsin-dong Doosan Apts, Yangcheon-gu Byucksan blooming Apts and Dongdaemun-gu Je-gi dong Isu Brown Apts have done similar activities. These resident organizations received financial aid from Seoul and set a goal to reduce energy consumption between 5% and 15% in addition to replacement with LEDs, energy-saving education classes, and electricity-saving competitions. They have their residents take part in energy production through the installation of mini-sized solar panels. With the aim of installing mini-sized solar panels, Je-gi dong Isu Brown Apts have carried out the project. Movement of Energy Self-Sufficient Village at the Apts level contributed to increasing electricity saving through installation of mini-sized solar panels. According to Seoul Metropolitan Government, Apts selected as Energy Self-Sufficient Village indicated 8–9% of energy saving on average (Seoul Metropolitan Government 2015).

After the Fukushima accident, various movements of Energy Self-Sufficient Village led by Seoul are based on advanced civic awareness and the 'one less nuclear power plant' policy. Shown in the case of Sungdaegol people and Kwanak Resident Network, there was the policy of Seoul that was a driving force to cultivate residents' recognition about energy transition and related activities. That is because the government which funded the projects such as Energy Self-Sufficient Village brought the expansion of residents' participation and energy education to improve the energy awareness. Energy movements led at the level of Apts including Kang Dong gu case was initiated from resident organizations aiming to utilize Seoul's policy rather than from the recognition of energy transition. However, as Seoul's energy policy suggested the direction of the movements, they were more passive than Sungdaegol people, Sungdaegol people and Kwanak Resident Network having discussed cooperative establishment beyond energy-saving plants, Apts residents' organizations funded by Seoul remained sluggish in the installation of mini-sized solar panels. And the term, 'Energy transition' is rarely used in activities of Apts organizations. Since the Fukushima accident, residents' spontaneous involvement in the Energy Self-Sufficient Village movement has represented 'energy citizen' engaging in energy production participation like cooperatives including electricity saving.

Conclusion

Civil society in Korea started considering energy issues, opposing the unilateral construction plan of nuclear waste disposal facilities. Until late 1980, nuclear energy policy including nuclear waste disposal was the core of energy policy, and the only the government decided its content. Civil society initiatives in Korea experienced advanced civic consciousness through the pro-democracy movement in the 1980s resisted government policy. This resistance was against nuclear energy policy, and regional residents and civil organizations formed countermeasure committees nationally to prevent the government from unilaterally selecting and constructing nuclear waste disposal facilities. This civil society movement gave rise to delaying the policy enforcement and introducing civil participation like the referendum system in the process of selecting sites. Though ritualistic, this movement brought some change in the policy, which was not carried over into the movement against nuclear power plant construction.

The perspective on energy transition with the aim of changing the energy system was issued, and an alternative energy system began to establish in 2000. Civil Centres for Alternative Energy took the lead in the movement of civil power plant construction by using small solar power plants for the first time in Korea. Those actions showed the appearance of a civil energy movement that was different from the traditional anti-movement. In the new grassroots energy movement, citizens could make an alternative to nuclear energy by themselves. As small power plants fuelled by renewable energy sources gathered momentum by introducing Feed in Tariffs in 2002 and YMCA or cooperatives joining in the building of civil power plants, 'civil power plant' construction went across the board. Renewable power plant construction is regarded as the democratization of the energy system, considering that energy production system monopolized by the government or conglomerates was transformed by civil participation.

With the foundation of civil power plants, Deungnyong village and Yeondaedo attempted to convert the energy system at the local level. Energy-saving as well as transformation plans like the construction of renewable energy facilities could be implemented in cooperation between local and national NGOs, local governments, and local activists. With regards to residents' participation, two cases presented some problems in the grassroots movement for energy transition at local level. Most of Deungnyong village residents took part in NGO activity passively, and Yeondaedo residents also did not engage directly in the renewable energy facility investment. Rather, photovoltaic power plants of Namjeon-ri and Jejudo are considered to contribute 'energy citizen' formation through the construction of plants owned by communities.

The civil energy movement and the grassroots movement led by civil groups, not by civil organizations, started after the Fukushima accident. The concept of Post-nuke was deployed in society, and photovoltaic power plant cooperatives in which minority investors joined began to be organized nationally. Not only was energy transition attempted but energy saving including the installation of mini-sized photovoltaic panel was realized actively at the village and Apts level. 'Sungdaegol people' is a good example that shows development of the grassroots movement from energy-saving movements to the foundation of cooperatives. The 'energy citizen' that Devine-Wright defined can be found in these activities. However, the deployment of a grassroots movement is restricted by the strict government policy addicted to nuclear energy. The inefficient support policy of renewable energy, such as closure of FIT, has hindered renewable energy cooperative from vitalizing. In conclusion attempts of individual citizens or citizen groups to energy transition are not meaningful without any appropriate government policy. The growing citizen group engaged in energy transition in Seoul showed that the policy for 'one less nuclear power plant' promoted the citizen movement. The energy transition in Korea requires the institutionalization of transition policy along with civil participation in the energy system through the grassroots movement.

Notes

1 'gun' is an administrative division. South Korea is divided into eight provinces (do), which are subdivided into a variety of smaller entities, including cities (si), counties (gun), districts (gu), towns (eup), neighborhoods (dong), and villages (ri).

2 Seoul defined energy self-sufficient village as high energy self-sufficient Village by saving energy, increasing efficiency, and minimizing received energy from exterior facilities by self-production with the help of new renewable energy sources (Seoul Metropolitan Government 2012).

References

CEA (2000). The Prospectus of Center for Energy Alternative, Retrieved 10 December 2016, from http://energyvision.org/.

Chang, Y. B. and J. K. Han (2008). *Towards participatory science and technology policy Making in Korea*. Seoul: STEPI.

Cho, M. S. and S. J. Yun (2016). Shaping Process of ecological citizenship by small group activities for energy self-sufficient village. Proceedings from the Annual Conference of Korean Association for Environmental Sociology, 1–26, Gongju.

Devine-Wright, H. and P. Devine-Wright (2005). Representing the demand side: 'deficit' beliefs about domestic electricity users, Retrieved 10 October 2016, from http://www.eceee.org/library/conference_proceedings/eceee_Summer_Studies/2005c/Panel_6/6163devine_wright/paper.

Devine-Wright, P. (2007). Energy citizenship: Psychological aspects of evolution in sustainable energy technologies. In J. Murphy & A. Sterling (Eds.), *Governing technology for sustainability for sustainability* (pp. 63–86). London: Earthscan,.

Dobson, A. (1998) Ecological citizenship: a disruptive influence? Retrieved 10 October 2016, from http://www.vedegylet.hu/okopolitika/Dobson%20-%20Ecological%20Citizenship.pdf.

Dobson, A. (2007). Environmental citizenship: Towards sustainable development. *Sustainable Development*, 15, 276–285.

Flyn, R., P. Bellaby and M. Ricci (2008). Environmental citizenship and public attitudes to hydrogen energy technologies. *Environmental Politics*, 17(5), 766–783.

Green Korea (2009). *Village who dreams an energy self-sufficient village. Making Buan Deungyoung energy self-sufficient*. Seoul: Green Korea.

Hong, D. H. and Y. H. Lee (2014). Energy movements and the typology of energy citizenship in Korea. *ECO*, 18(1), 7–44.

Hong, D. H. (2016). The promotion and transformation of nuclear industry in Korea (PhD dissertation, Seoul National University).

Jin, J. Y. (1991, November 6). Building the national network of anti-nuclear power plant and waste disposal facilities. *The Hankyoreh*.

Kang, J. Y. and T. D. Lee (2016). The role of intermediary supporting organization in energy regime change: Comparative case studies of energy self-sufficient islands in Korea. *Space and Environment*, 26(1), 139–176.

KEEI (2015). *Yearbook of Energy Statistics*. Korea Energy Economics Institute.

Korean Federation for Environmental Movement (KFEM) (1993). Codes of Korean Federation for Environmental Movements. in Institute of Mosim and Salim (ed.). *Collection of Documents on Life Movements*. Unpublished material.

Koo, M. G. (2013). South Korea's feed-in tariff program and its implications for new industrial policy: Policy design, implementation, and learning. *The Korean Journal of Public Administration*, 22(1), 1–27.

Leach, M. and I. Scoones (2005). Science and citizenship in a global context. in M. Leach, I. Scoones, and B. Wynne (eds). *Science and citizens: Globalization and the challenge of engagement* (pp. 15–40). New York: Zed Books.

Lee, H. S. (2016). Fukushima five years, from Anti-Nuke to Post-Nuke. *Green Review*, 147, 100–122.

Lee, J. P. and J. K. Han (2014). Implication of community energy and energy citizenship in studies of energy transition. Proceedings from the Annual Conference of Korean Association for Environmental Sociology (pp. 79–107). Seoul.

Lee, Y. H. (2011). *Science, technology and democracy*. Seoul: Moonji Publishing.

Lee, Y. J. (2014). An analysis of energy self-sufficient village for transition of local energy. Proceedings from the Annual Conference of Korean Association for Environmental Sociology (pp. 118–144). Seoul.

Lee, Y. J. (2016). The niche strategy of civilian-led energy self-sufficient village for the energy transition – The Case of SUNGDAEGOL Village in Seoul. (PhD Dissertation, Seoul National University).

Marshall, T. H. (2009). Citizenship and social class, In J. Manza & M. Sauder (eds), *Inequality and society* (pp. 148–154). New York: W.W. Norton and Co.

Melo-Escrihuela, C. (2008). Promoting ecological citizenship: Rights, duties and political agency. *ACME: An International E-Journal for Critical Geographies*, 7(2), 113–134.

MOTI (2014). *The Second National Energy Plan* (Ministry of Trade, Industry and Energy).

Park, J. H. (2008). System transition, social discourse about climate change, and renewable energy. *Environmental Philosophy*, 7, 99–136.

Park, J. H. (2015). State and tasks of energy cooperatives in Korea: From the perspective of energy citizenship. *ECO*, 19(1), 173–211.

Seoul Metropolitan Government (2015). *Manual for Building Energy Self-sufficient Village*. Seoul: SMG.

Yun, S. J. and H. Y. Shim (2015). Possibilities and institutional limits of citizens' solar power cooperatives as a strategic niche for energy transition: Focusing on the case of Seoul. *Space and Environment*, 25(1), 140–178.

Yu, U. S. (1989, August 10). Urgent investigation of nuclear power plant. *The Dongailbo*.

Part III

Energy and social risk

7 ‘Structural disaster’ beyond Fukushima

Messages from the sociology of science and technology

Miwao Matsumoto and Kohta Juraku

1 Introduction

The Fukushima nuclear power plant accident was extremely shocking. Many have argued that we should learn from the disaster and improve our technology, institutional frameworks, organizational cultures, and many other aspects of society to prevent another tragedy involving advanced science and technology. Indeed, many practical lessons and recommendations have been itemized in various fields. Some of them suggested technical improvements to nuclear power plants. Others identified problems with organizational behaviours and possible remedies. There has also been a series of ethical reflections that place the blame on negligent nuclear experts. Apart from these, we call attention here to different perspectives to grasp the mechanism behind the accident, the mechanism which has contributed to reproducing serious negative incidents with devastating consequences in the complex science-technology-society interface. If the mechanism continues to work, similar patterns should be repeated in different times and places despite the many ‘improvements’ that were supposedly made.

For example, Matsumoto (2012; 2013; forthcoming) critically analysed the devastating failure in transmitting critical information about the accident to the people when the Japanese government faced unexpected and serious events after 11 March 2011. Secrecy toward outsiders generated this failure; secrecy toward the people who were forced to evacuate from their birthplaces, toward the people who wanted to evacuate their children, and others. Matsumoto attempted to shed fresh light on the structural causes of the Fukushima accident by illuminating the patterns of behaviour by the agents involved in the accident and highlighted the workings of ‘institutionalized secrecy’ as a new dimension of the problem. He then extended the sociological implications of ‘institutionalized secrecy’ within a comparative perspective with a little-known but serious naval accident that occurred immediately before World War II, focusing on the subtle relationship between success and failure in the complex science-technology-society interface. To contextualize the similarities and differences between individual ‘institutionalized secrecy’, an integrating conceptual framework for in-depth descriptions and analyses is required. The concept of ‘structural disaster’ was introduced for that purpose (Matsumoto 2002/2012; 2012; 2013; forthcoming). The concept is intended to play the role of integrating framework and delve into the deep layers of the situation before, during, and after Fukushima.

Also, Juraku (2013a) revealed the historically shaped structure that made the consequences of the Fukushima nuclear accident worse – the centralized multi-unit siting strategy – that used to be considered an important factor that enabled rapid and steady growth of the Japanese nuclear power plant fleet by promoters of the national nuclear programme before the 3.11 earthquake-tsunami-nuclear disaster. By tracing the history of delicate but serious social struggles that have continued since the early stages of Japanese nuclear siting, he showed the existence of the stubbornly similar failure trajectory that also amplified the damage of the 3.11 disaster.

In this chapter, from these perspectives, we will critically examine another important trajectory of nuclear policy: high-level radioactive waste (hereafter, HLW) management policy in Japan. HLW management is an extreme example of a high-uncertainty problem in the science-technology-society interface and thus it requires well-articulated and elaborated policy measures to handle it appropriately (Juraku 2013b; 2015; 2016). Due to the very long time frame and global scale of the issues involved, HLW management calls for carefully arranged, and socially reasonable 'allocation of almost infinite responsibility' (Matsumoto 2009; 2012).

After more than 10 years of siting stagnation and the impact of the Fukushima nuclear accident, the Ministry of Economy, Trade and Industry (hereafter, METI) of the Japanese government initiated policy reform for its HLW management in 2013. As a result of almost two years' discussion by the expert committees of METI,[1] the revised 'basic policy for final disposal' was decided by the cabinet in May 2015 (METI 2015a). This reformed policy recognizes the high degree of uncertainty involved in the programme and admits a lack of public support. The reformed policy incorporates almost all of the recent international trends in HLW management, such as 'reversibility of management programme and retrievability of disposed nuclear waste' (hereafter, R&R). At first glance, it seems that the failure trajectory of Japanese HLW management policy has been corrected. However, we argue that the process shows essential symptoms of 'structural disaster', which will be explained in more detail in the following sections. In particular, the reform process itself still incorporates a series of wrong-headed traditions for nuclear policy and decision-making processes, so that the process could subsequently reproduce inappropriate and unreasonable policy choices.

2 Background of the recent HLW policy reform in Japan

2.1 Long stagnation of the open solicitation siting process

As mentioned above, Japanese HLW management policy has suffered from over 10 years of siting stagnation since the Nuclear Waste Management Organization of Japan (hereafter, NUMO) was established in 2000 by the Designated Radioactive Waste Final Disposal Act, which was enacted in the same year. Only one municipality has applied for NUMO's open solicitation process for candidate areas for geological final disposal facilities; a process that was introduced in 2002. Toyo-cho town in Kochi Prefecture applied in 2006–2007, but the application was cancelled during the initial stage of the process due to strong opposition from the residents. It caused very serious local disputes and social conflicts within the town and surrounding area. As a result, Mayor Yasuoki Tashima was obliged to resign (Tashima 2008, Sugawara and Juraku 2010, Saigo *et al.* 2010).

After that, no applicants came forward and contacted NUMO, though METI added 'invitation by the government' to their site selection process in 2008 and NUMO tried to strengthen their public relations campaign via TV advertisements, public symposium and so on. In addition, the impact of the Fukushima nuclear accident in 2011 caused the virtual suspension of the HLW siting process.

Consequently, METI and NUMO went over 15 years without any candidate municipality, contrary to their optimism about siting when they set out to establish the HLW disposal programme in 2000.

2.2 Science Council of Japan's report and criticism

In September 2012, the Science Council of Japan (hereafter, SCJ) issued its report on HLW management and disposal. This was a report responding to a request for review made in September 2010 by the Atomic Energy Commission of Japan (hereafter, AECJ). The report was originally scheduled to be issued after a one-year discussion at the special committee on this subject in the SCJ, but this schedule was abandoned due to the Great East Japan Earthquake and subsequent Fukushima nuclear accident in March 2011.

The SCJ commission's report included a critical review that was far beyond the AECJ's original request. It asked the SCJ to 'supply an explanation and information for the public on the HLW disposal programme' (Science Council of Japan 2012). It recommended the following steps for national HLW management policy:

1 Fundamental reform of the HLW management/disposal policy.
2 Recognition of the limits of scientific/technical capacity and the importance of scientific autonomy.
3 Restructuring of the policy framework backboned by 'interim storage' and 'total volume control' of HLW.
4 Development of a convincing, fair decision-making process.
5 Stepwise consensus building in the arena for participatory discussion.
6 Recognition of long-term, persistent efforts to find a final answer to the problem.

In particular, the report presented a very critical view of the short-term feasibility of geological disposal in Japan. It presented two reasons for this: firstly, there is a lack of public trust in and scientific/technical (especially geological) knowledge for the final disposal programme. The SCJ commission argued that the current scientific/technical expertise is insufficient to convince society of the validity of long-term geological disposal, not only because nobody anticipated such a mega-quake and subsequent huge tsunami, as observed in the Fukushima accident, but also because nuclear power stations were not made resistant to the impact of such a disaster. Secondly, the report pointed out that the Fukushima nuclear disaster totally destroyed public trust in nuclear science, technology, and policy. It went on to state that there is an urgent need for a fundamental reform of national nuclear policy with the understanding that HLW disposal is an indispensable part of such a reform.

The report never denied the feasibility and advantages of geological disposal *in general*. However, it argued that it would be infeasible *in Japan* for the two reasons mentioned above. The report criticized the efforts made by METI, NUMO, and other related agencies as 'procedurally inverted practices', because they tried to trivialize the HLW management issue as an ordinary siting problem and had never tried to achieve an explicit national consensus over the Japanese nuclear programme, notwithstanding that HLW management could affect the national consensus as a whole.

Thus, based on the condemnation, the report suggested a zero-scratch discussion on the direction of nuclear utilization and HLW management. To realize this, it recommended 'interim storage' and 'total volume control' of HLW until a new policy is established and more permanent HLW management options become available. The report said that the 'interim storage' would last for several decades or even centuries. These critical arguments attracted fairly wide public attention, probably because the post-Fukushima public opinion reflected the importance and difficulty of the so-called back-end issues of nuclear utilization in the wide-ranging discussions triggered by the impact of the Fukushima disaster.

The AECJ replied to the SCJ report in December 2012. They declined to accept the fundamental reform and cancellation of the national geological disposal programme, defending the validity and appropriateness of the current scientific/technical expertise and the legal and procedural legitimacy of the institutionalization process of the programme. However, they had to admit the importance and need to make the programme more participatory and flexible; therefore, they encouraged the related agencies to take actions for these reforms (Atomic Energy Commission of Japan 2012).

2.3 METI's policy reform since 2013

Responding to such a situation and facing a critical social stance toward nuclear-related issues in the post-Fukushima Japan, METI decided to start the discussions for their policy reform of HLW management again in its own right. In May 2013, METI revived the 'Radioactive Waste Working Group' of the Advisory Committee on Energy and Natural Resources as an expert committee to consult on the reform. They also established the 'Geological Disposal Technology Working Group' as a sister commission to the 'Radioactive Waste Working Group' in October 2013.

Both of these working groups published their interim reports in May 2014 (METI 2014a; 2014b). After some follow-up discussion in those working groups, and METI's in-house policy decisions at 'ministers' meetings for final disposal' held in December 2013, September 2014, and May 2015, the revised 'basic policy for final disposal' was eventually decided by the cabinet on 22 May 2015 (METI 2015a).

This reformed policy includes the following points:

1 Clear recognition of uncertainty and the lack of public support and emphasis on trust re-building through more participatory and transparent policy and practices in the HLW final disposal programme.
2 Explicit adoption of R&R (reversibility and retrievability), especially to reserve the future generations' options of policy and technology alternatives.
3 Promoting R&D such as spent fuel direct disposal, partitioning, and transmutation to make alternative options more realistic.
4 Introduction of suitable area screening before the official start of the siting process to avoid political conflicts at a too-early stage, to ensure scientifically sound and reasonable siting.
5 Recognition of the importance of further spent fuel and vitrified HLW storage capacity to keep some flexibility in the HLW management policy.
6 Introduction of a local participatory dialogue process in future candidate areas to make the process more open, transparent, and participatory.
7 Acknowledgement of the need for continuous review of the technical reliability of geological disposal technology, due to the particular geological characteristics of Japan.
8 Enforcement of the independent oversight of HLW disposal policy and practices.
9 Reform and improvement of NUMO's corporate governance and empowerment of its capacity as the implementer.
10 Encouragement of social science studies on HLW management and disposal issues.

Almost all of the revisions are in line with global trends in HLW management policy as far as their general direction is concerned.

3 Review and discussion of the reformed policy: problems in the relationships between expertise and policy

3.1 Under-elaborated policy concepts

However, the above list is still under-developed in terms of its technical, practical, and ethical considerations as a concrete policy package. The reformed 'basic policy for final disposal' simply shows the HLW disposal promoters' commitment to the contemporary international standards of participatory, stepwise, transparent, and comprehensive policy strategy and practice. There are many things to be deliberated, elaborated, contextualized, and fixed before materialization.

For example, it is still open to discussion how to decide the scope, condition, and methods of R&R concepts. No careful evaluation of the trade-offs centring around R&R has been done and this fundamental question has been postponed until the next round of HLW policy restructuring in Japan. Costs; safety; security; risks; ethical, legal, and institutional problems; consistency with regulation; and many other concrete and detailed discussions related to R&R have not yet been held. In addition, the design and scheme of the participatory local dialogue process and the independent oversight and review processes are to be discussed and decided from now. These items also contain many trade-offs between the flexibility, legitimacy, stability, and efficiency of the policy.

All of these items must be examined carefully and answered with clarity to realize them as an effective and sound approach to the better management of HLW. They will then be left to ex post assessment by all of us facing the post-Fukushima situation.

3.2 Discovered, or re-discovered: missing links with past and international discussions

In this connection, it should be noted that the policy reform items listed above are not something new to international practical and academic efforts in the field of HLW management. To the contrary, they are very common to experts in the relevant technical and policy fields: in fact, the relevant experts have already discussed, tried, developed, and elaborated on them for decades. Therefore, the introduction of those policy items themselves is not epoch-making or innovative. By the same token, many reports relevant to the policy items have already been published by international organizations such as the OECD/NEA and the National Academy of Science in the US (National Academy of Science 1990; 2001; 2003, OECD/NEA 1977; 1995; 2011). To be fair on the originality, the Japanese governmental sector, and probably the academic sector (including the SCJ), just 're-visited' or 're-discovered' these ideas.

Of course, this would not necessarily be criticized if they could make use of the knowledge, know-how, and lessons obtained through such reports. In that case, latecomers' advantages could be expected in the process of learning lessons from the accumulated policy trajectories. However, this is not the case because few of them were taken up for intensive consideration and are still under discussion in the upcoming METI policy process. The 'basic policy for final disposal' mentions little about the concrete method of implementation of those concepts as suggested above. That is to say, there exists a very high degree of 'type 2 underdetermination' (Matsumoto 2010),[2] so that whether the above-mentioned government-led policy reform leads to public good heavily depends on its method of implementation and actual institutional design.

In assessing the prospect of policy reform, one should note that the needs for introducing R&R, a participatory stepwise process, and transparent policy practices were pointed out even in the important report of an AECJ subcommittee (*Kou-reberu Housha-sei Haikibutsu Syobun Kondan-kai*, the Special Committee on

Disposal of High-Level Radioactive Waste), in the late 1990s, before the Japanese HLW disposal programme was established (Atomic Energy Commission of Japan 1998). It is true that those items were not explicitly acknowledged nor fully implemented under the previous Japanese HLW policy. However, the basic framework of the AECJ subcommittee report was employed by the government at the time while the need to introduce R&R, a participatory stepwise process, and transparent policy practices were acknowledged but not employed. Thus, selective negligence has played a significant role in the decision-making processes of METI, which has been responsible for both the siting of nuclear power stations and HLW disposal.

Regarding the proposal by the SCJ commission on HWL disposal made public to correct the trajectory of HLW policy, it was rejected during the governmental policy reform process precisely because of the selective negligence of the governmental sector. The SCJ commission consisted of non-nuclear-related academics to make the study more neutral and free from the intervention of nuclear interest networks. The members of the commission were both natural and social scientists, and a social scientist chaired it. The importance and effectiveness of such an interdisciplinary study group has long been noted. Unfortunately, as far as this commission was concerned, the reality seemed to be different from the expectations. It is true that the social scientists in the commission had a sharp awareness of public issues. However, what is striking is the fact that most of them lacked relevant expertise indispensable to a proper understanding of the meanings and social importance of issues such as HLW disposal. They also lacked awareness of the irrelevance of the lacking expertise, be it technical or social scientific, as the members of the commission who were supposed to represent the Japanese academic sector.[3] In materializing interdisciplinary studies, both the possibility of 'fruitful heterogeneous collaboration' and that of 'mutual impoverishment' are equally left open depending on the working milieu of the studies (Matsumoto 2002/2012). One of the essential conditions making up such a working milieu is that technical matters are discussed by experts who have relevant expertise in the matter. To be accurate, the natural scientists on the commission had technical expertise in their own fields, but there existed no one who had full expertise in HLW disposal, not to speak of the social, ethical, and political meanings and importance of HLW disposal issues.[4]

As a result of this kind of institutionalized bad design in the mobilization of expertise for the public interest, the points made by the report resulted in what is called a very rough repeat of the general discourse from the past – a discourse that is general enough to have been made even by the members of the 'nuclear village', including nuclear waste experts, lagging far behind the current research front of related fields. Few agents criticized it as a kind of 'Columbus' re-discoveries' and questioned the quality of the report while it was highly popularized by journalists, activists, and some social scientists without substantially relevant peer reviews by those with sufficient and appropriate expertise in the various problems involved in the issue.

This situation is structurally similar to the selective negligence observed in the decision-making processes of METI mentioned above. From such a viewpoint, the structural similarity is running not only through experts, including social scientists, from the academic sector, but also through 'lay experts' from the governmental and industrial sectors who are engaged in the promotion of nuclear power generation by selectively dismissing discouraging information as 'misunderstandings'.[5] This situation means that the current institutional design surrounding the academic sector can self-negate themselves because the institutional design could thus contribute to skipping relevant peer reviews and selectively neglecting the latest expertise – at least since 1986, when the criterion of the membership of SCJ was changed to nomination by the prime minister from election. All of these even reinforce the rhetoric employed by other, non-academic sectors to legitimate status quo: using such rhetoric one can argue more strongly, 'we still need more public understanding of anything nuclear, because even the best scholars supposed to represent the academic sector did not know the elementary matters in HLW disposal'. This is only a possibility in the current configuration of the discourse revolving around HLW disposal issues in Japan, but it has a serious and general implication for all of us who are in the post-Fukushima era. That is to say, the rhetorical legitimation of a status quo of this kind can hollow out the expected social role of the academic sector.

3.3 Persistent continuation and expansion of the 'public relations' approach

Suffering from the structural failure of the policy reform process itself, Japanese HLW management policy has carried over its fundamental problem: the lack of explicit and due consent given by heterogeneous sectors of society to the principles on which measures to deal with HLW disposal are grounded. Before that, there lacks a socially shared awareness of the crucial importance of such a principle and the consent given to such a principle. Because HLW disposal is an extreme case of a highly compound problem in the science-technology-society interface accompanying infinite responsibility (Matsumoto 2009; 2012), HLW management policy essentially requires fundamental value judgment via democratic deliberation and decision making to make it legitimate and sound (Juraku 2013b; 2015; 2016).

However, as Sugawara and Juraku (2010) pointed out, the Japanese HLW programme had a tendency to trivialize the issue into a so-called 'siting problem', neglecting the fact that the core of HLW management is not just the siting of the facility, but the societal challenge of coping with unparalleled uncertainty. The authors argued that one of the reasons for this tendency was the past 'successful' experience of nuclear power expansion as represented by the many nuclear power stations sited in Japan. Policy-makers have long believed that the obstacles to their nuclear programme have little to do with technical expertise but are strongly related to so-called 'social acceptance' (political and social management of public trust and reputation). One of the most important foci of 'social acceptance' was the siting of nuclear facilities. This long-standing tradition seems to have weakened Japanese policy-makers' attention to the link between expertise, democratic decision-making, and the process of policy-making and implementation. They seem to be too much inclined toward the management of the political atmosphere and public opinion.

This tendency seems to be exactly followed by HLW management policy, unfortunately, because METI's 'new' policy is still totally in line with this long-standing tendency. After the policy reform process described in section 2.3, METI and NUMO began their collaborative campaign to encourage 'public understanding' of the reformed national HLW policy. They conducted 27 'nation-wide symposia' in big cities throughout Japan (nine in 2015, nine in 2016 and nine in the earlier half of 2017) to 'think together on this issue with our fellow citizens' (METI-NUMO 2016) and also provided a series of 'small group workshops', 'presentation deliveries', 'citizen dialogue meetings', and other public relation programmes, all of which are expected to serve as a showcase of participatory policy. The purpose of these activities was defined as 'to cultivate public understanding of the policy', according to METI brochures distributed at the 'Radioactive Waste Working Group' meetings.[6] METI and NUMO have eagerly carried out such events, gaining reflections from the working groups to improve 'communication', sometimes in very critical ways. In particular, they have tried to convey the intention and purpose of their brand-new 'introduction of suitable area screening' policy to the public during their two-year-long campaign.

However, the only result of the promotion of these seemingly participatory and virtually 'public relations' activities was negative political reactions from both local residents and local governments. From the early stage of METI-NUMO's campaign, not a few prefectural governors showed their refusal to participate in an area survey or host an HLW disposal facility, based on a nomination process carried out in a top-down manner. According to Kyodo Press's questionnaire survey that was carried out from late October to early November, 2015, 13 out of 47 prefectures in Japan expressed their refusal against the siting of HLW disposal facility and another eight prefectures answered their negative attitude to this issue. There was no prefecture showed positive attitude towards the siting in their survey result (Kyodo Press 2015).

Also, both METI and NUMO admitted that several participants of the 'nation-wide symposia' expressed their questioning of and/or criticisms against national nuclear policy as a whole, not just about the siting process for radioactive waste disposal (for example, METI 2015b). This could not be answered under the framing of the 'sitingnized' public relations activities. It should and could only be discussed through an open-ended social decision-making process (not public relations events) on national nuclear policy. In reality, however, METI's answer was 'to provide further information about national nuclear policy' (METI 2016).

Furthermore, NUMO's own public poll results showed that the majority of people have not changed their sceptical attitude toward the reliability of geological disposal, the availability of suitable areas for disposal facilities, and the geological disposal programme itself (NUMO 2016).

These facts strongly suggest that METI-NUMO's 'public relations'-oriented strategy has not been effective at promoting their own HLW disposal programme. Instead, it seems to have reinforced public distrust and local opposition to it. But they are continuing these 'public relations' events one after another, like a knee-jerk reaction. Occasionally they even suggest some practical improvements. Among other things, what is striking here is the fact that they keep reproducing substantially the same 'public relations'-oriented strategy with different appearances without any prospects for the future, probably because there is no substantial plan to trigger social deliberation and consensus-building on basic principles we can rely on. This state reminds one of the occurrence of a 'structural disaster'.

3.4 Problems with the Japanese policy formation process: signs of a 'structural disaster'

Based on the above, there must be a serious problem with the linkages between expertise, democratic decision-making, and the institutional design involved in the entire process of policy-making, implementation, and evaluation. This state should be recognized as a structural problem, not a sporadic or incidental one that can be ascribed to the personalities of the particular individuals concerned, the external situation, coincidences, or any other factors other than structural causes. We need to explore the nature and working mechanism of the structural cause to break through the institutional design that functions to reproduce the same policy failure repeatedly.

In this connection, we propose understanding this phenomenon through the concept of a 'structural disaster of the science-technology-society interface' (Matsumoto 2002/2012; 2012; 2013; forthcoming). The concept is designed to shed fresh light on the mechanism behind this persistently erroneous trajectory. According to the concept, it is possible that these persistent errors are not caused by some failure of science, technology, or society as separate entities. It could and should be considered as 'the failure of the science-technology-society interface' (Matsumoto 2012; 2013).

According to Matsumoto, a 'structural disaster' consists of one or more of the following five elements (Matsumoto 2012; 2013):

1 Following erroneous precedents carries over problems and reproduces them.
2 The complexity of a system under consideration and the interdependence of its units aggravate problems.
3 Invisible norms of informal groups virtually hollow out formal norms.
4 Patching over problems at hand invites another patching over as a temporary countermeasure.
5 Secrecy develops across different sectors and blurs the locus of agents responsible for the problems in question.

The absence of an appropriate link between expertise and policy can be brought about by some of these elements and their combination. For example, the long-standing custom of incomplete and unsystematic reference to some expertise and selective negligence of other relevant expertise was carried over from the pre-Fukushima nuclear policy making system in Japan and authorized (elements 1 and 3). Ad-hoc policy revisions were made instead of a fundamental reform of the entire nuclear policy (element 4). Responsibility for decisions and criteria of success and failure for policy measures were obscured and/or undisclosed (element

5). Persistent expansion of 'public relations'-oriented policy can also be interpreted as a result of a carry-over from the siting strategy of nuclear power stations coupled with the tradition of METI's advisory group steering know-how (all of the elements listed above are found here).

If we adopt this perspective, it follows that no single technical failure, obvious scientific misunderstanding, or single person is responsible for the continuation of such a devastating situation even after the Fukushima nuclear accident. For example, as mentioned above, the mobilization of scholars without expertise relevant to the problems at hand in the academic sector and the repeated reproduction of the 'public relations'-oriented policy in the governmental sector have been reinforcing each other, decreasing the possibility of drastic structural reform. The interface between those heterogeneous sectors of society suffers from serious structural problems. Illustrating this in terms of the allocation of due social responsibility, a 'structural disaster' does not mean the exemption of particular agents from responsibility for the disaster; on the contrary, it means that everyone should take due social responsibility and there is no exception to this rule. This perspective suggests that the problems with Japanese HLW management policy and practices are not solvable via piecemeal technical improvement nor by reform based on some 'cookbook'; rather, a redesign of the institutional structure of the interface is required.

4 Concluding remarks

Most of us now know that the overly strong inclination toward 'social acceptance', which was pointed out in the previous section, is not sound and is no longer legitimate. We know this because we witnessed the Fukushima nuclear accident, which was undoubtedly caused by technical insufficiencies coupled with institutional barriers, not by a lack of 'social acceptance'. This insight is one example of what the concept of 'structural disaster' reveals about the Fukushima nuclear accident.

We need to re-design the links between the policies and the past and international experiences to re-balance the relevant expertise and public interest. One of the most effective and valid bridges between the two is the redesign of social arrangements to properly utilize relevant expertise. Academic or non-academic expertise, technological or non-technological expertise, domestically accumulated expertise or internationally shared expertise all tell us the locus where appearance begins to diverge from reality. Whether we are in an emergency situation or not, expertise should be referred, interpreted, discussed, and implemented in a substantially fair manner in the policy-making process.

To materialize this, it is important to implement a participatory, open, and interdisciplinary arena in the policy and decision-making processes to encourage interactions between policy-makers, practitioners, technical experts, social scientists, stakeholders, and citizens. All of them have different kinds of knowledge and expertise, and their blind spots could be supplemented by others, leading to better policy design and practice that learns bitter lessons from the past. Here, it is crucial to put aside anti-intellectualism in a broad sense because that is one of the easiest things to assimilate and diffuse in a creeping manner from one sector to another throughout the world. Anti-intellectualism in a broad sense here means two things: first, hollowing out the working norms of the academic sector that have guaranteed the quality of expertise; and second, quasi-expertise resulting from such a hollowing out could be mobilized in the public sphere to perpetuate the status quo of the science-technology-society interface, even after devastating events such as the Fukushima nuclear accident.

The participatory process has tended to be considered as a method of 'public relations' to earn 'public acceptance'. After the devastation in Fukushima, however, it has been thoroughly made clear that such an understanding is partial and practically ineffective. But this is not so because of the intervention of the 'nuclear village'. What is said is that such an understanding is partial and ineffective because it assumes inequality in accepting the social 'bad' between the citizen sector and other sectors of society. We should reconsider the needs, methods, and goals of such a method of policy-making, as well as the implementation and decision-making processes, and move away from past tribalism as exemplified by 'nuclear village' and other 'villages' to a fairer system to allocate social responsibility for expertise, democracy, and policy-choice. We hope that many social scientists throughout the world will participate in this new endeavour afresh.

Acknowledgement

Part of this research was supported by JSPS KAKENHI Grant Number 26750096.

Notes

1 One of the authors, Kohta Juraku, participated in this policy reform process as a member of 'Radioactive Waste Working Group' of the Advisory Committee on Energy and Natural Resources. Also, he worked for another commission established by the Science Council of Japan (mentioned in section 2.2) as a research associate.

2 According to Matsumoto (2010), 'If there is really a possibility of underdetermination of scientific knowledge, then it is obvious that people are all the more likely to see the underdetermination of the entire process of policy making, implementation, and evaluation in the science-technology-society interface. This is not only because indeterminate scientific knowledge is one of the key elements involved in the entire process of policy making, implementation, and evaluation, but also because the process includes a different kind of underdetermination of its own. This is due to parameters of a different nature, such as the delicate relationship between experts and the general public, the changing limit of budgets and the necessity of its execution within a strict date limit, the need for face-saving in the public sphere, the particularistic interests of a specific public body, and others, which are usually fixed by various implicit assumptions'. Based on this understandings, he calls 'the underdetermination of scientific knowledge "type 1" underdetermination, and all other underdeterminations involved in the process of policy making, implementation, and evaluation "type 2" underdetermination.'

3 The above descriptions are based on the specialities of those members of the commission put on the list of members of the commission, and its minutes. See Science Council of Japan (2012).

4 It is surprising to find that the SCJ report failed to meet the critical requirements listed below to be academically valid and socially/politically effective (Juraku 2013b):

1 Little reference to previous reports, papers, and other expertise.
2 No careful discussion about the terminology of the report (ex. novelty of the concept of 'interim storage').
3 No members with a background in geological disposal in the commission.
4 No systematic selection of reference and witnesses.
5 No evidence provided for important arguments.
6 Changing the original question given to the commission without the consent of the AECJ.

5 Typical examples of 'misunderstanding' by the SCJ commission are as follows: 'geological disposal is irreversible and irretrievable in principle', 'retrievability was absolutely lost when we closed the underground facility', and 'reversibility is bureaucrats' translation of retrievability and it has limited flexibility compared to retrievability'. All of the above statements are simply wrong from a technical perspective. This could be understood as a case of 'technological populism' (Collins and Evans 2007) *by the academy* and the rejection of it by the promoters' technocracy.

6 All of the distributed materials are available online on METI's website (in Japanese): http://www.meti.go.jp/committee/gizi_8/21.html

References

Atomic Energy Commission of Japan (1998). Ko-reberu Hosha-sei Haiki-butsu Syobun ni muketeno Kihon-teki Kangaekata nitsuite [Basic Framework for Final Disposal of High-level Radioactive Waste]. (in Japanese)

Atomic Energy Commission of Japan (2012). Kongo-no Ko-reberu Hosha-sei Haikibutsu no Chisou-shobun ni kakaru Torikumi ni-tsuite [Statement: On the Future Measures for Geological Disposal of High-level Radioactive Waste], December 2012. (in Japanese)

Collins, H. and R. Evans (2007). *Rethinking expertise*, Chicago, IL: University of Chicago Press.

Juraku, K. (2013a). The failure of Japan's 'successful' nuclear programme: Structural problems revealed by the Fukushima Daiichi nuclear accident. In R. Hindmarsh (ed.), *Nuclear disaster at Fukushima Daiichi: Social, political and environmental issues*, pp. 41–56, London: Routledge.

Juraku, K. (2013b) Ko-reberu Hosha-sei Haiki-butsu Syobun no 'Muzukashisa' e-no Taisho no Michi-suji wo Saguru: Motome-rareru Chi-no Shakai-teki na Kyou-yu to 'Kachi-Sentaku' no Giron [To deal with the 'difficulties' of high-level radioactive waste management: the necessity of 'value judgment' discussions and knowledge sharing in the society]. *Kagaku*, 83(10), pp. 1164–1173. (in Japanese)

Juraku, K. (2015) Ko-reberu Hosha-sei Haiki-butsu Syobun ni-okeru 'Anzen' no 'Muzukashisa' wo-megutte: Nihon-gakujutsu-kaigi to Keiza-sangyo-sho ni-okeru Saikin-no Giron to sono-Gan-i ['Difficult' issues centring on 'safety' of high-level radioactive waste management: A commentary on the recent discussions at Science Council of Japan and METI]. *Kagaku*, 85(3), pp. 307–313. (in Japanese)

Juraku, K. (2016) Kou-reberu Housha-sei Haiki-butsu Syobun no 'Ricchi-mondai-ka' no Mondai-ten: Saikin-no Seifu-no Seisaku-minaoshi to Kongo-no Academy no Yakuwari-wo Megutte [Problems of 'sitingnization' of high-level radioactive waste disposal:

Recent governmental policy reform and role of the Academy]. *Trends in The Sciences*, 21 (6), Japan Science Support Foundation. (in Japanese)

Kyodo Press (2015) Kakuno-gomi 21 Do-fu-ken Kyohi Shobun-chi Ukeire [Nuclear waste: 21 prefectures refuse to host the disposal site], 29 November 2015. (in Japanese)

Matsumoto, M. (2002=2012) *Chi no Shippai to Shakai: Kagaku-Gijutsu wa Naze Shakai ni Totte Mondai-ka* [The structural failure of the science–technology–society interface]. Tokyo: Iwanami Shoten (in Japanese)

Matsumoto, M. (2009) *Techno-Science Risk to Shakai-gaku: Kagaku-Shakai-gaku no Arata-na Tenkai* [Techno-science risk and sociology: New perspectives of sociology of science and technology], Tokyo: The University of Tokyo Press. (in Japanese)

Matsumoto, M. (2010) Theoretical challenges for the current sociology of science and technology: A prospect for its future development. *East Asian Science, Technology and Society: an International Journal*, 4, 129–136.

Matsumoto, M. (2012) *Kozosai: Kagaku-Gijustu-Shakai ni Hisomu-Kiki* [The structural disaster: Crisis hidden in techno-scientific society]. Tokyo: Iwanami Shoten (in Japanese)

Matsumoto, M. (2013) 'Structural disaster' long before Fukushima: A hidden accident. *Development and Society*, 42(2), 165–190.

Matsumoto, M. (forthcoming) *The sociology of 'structural disaster'*. London: Routledge.

Ministry of Economy, Trade and Industry of Japanese Government (METI) (2014a), Hosha-sei Haiki-butsu WG Chukan Torimatome [Interim Report of Radioactive Waste WG], May 2014. (in Japanese)

Ministry of Economy, Trade and Industry of Japanese Government (METI) (2014b), Saishin-no Chiken-ni Motozuku Chiso-shobun Gujutsu-no Sai-hyoka: Chishitsu-kankyo Tokusei oyobi Chishitsu-kankyo-no Choki Antei-sei nitsuite [Reevaluation of geological disposal technology based on the latest scientific knowledge: On the characteristics of geological environment and long-term stability of geological environment], May 2014. (in Japanese)

Ministry of Economy, Trade and Industry of Japanese Government (METI) (2015a), Basic policy on the final disposal of designated radioactive wastes was revised: The government of Japan will play a proactive role in resolving the issue of designated radioactive wastes, Press release in 22 May 2015. http://www.meti.go.jp/english/press/2015/0522_01.html.

Ministry of Economy, Trade and Industry of Japanese Government (METI) (2015b) Kihon-hoshin Kaitei-go no Taiwa-katsudo to no Torikumi-Jokyo to Kongo no Hoko-sei: 'Kokumin Taiwa Gekkan no Furikaeri wo Chushin-ni' (Status and orientation of public relation activities after the amendment of the national basic policy: Towards the scientifically suitable area mapping) Document No. 1 of the 23rd meeting of Radioactive Waste Working Group, METI, 25 November 2015. http://www.meti.go.jp/committee/sougouenergy/denryoku_gas/genshiryoku/houshasei_haikibutsu_wg/pdf/023_01_00.pdf.

Ministry of Economy, Trade and Industry of Japanese Government (METI) (2016) Kagaku-teki Yubo-chi no Teiji ni Muketa Kongo no Torikumi oyobi Hon WG no Susume-kata (Measures towards the Introduction of Scientifically Suitable Areas and Agenda of this Working Group) Document No.1 of the 24th meeting of Radioactive Waste Working Group, METI, 27 January 2016. http://www.meti.go.jp/committee/sougouenergy/denryoku_gas/genshiryoku/houshasei_haikibutsu_wg/pdf/024_01_00.pdf.

National Academy of Science (1990), Rethinking high-level radioactive waste disposal: A position statement of the Board of Radioactive Waste Management.

National Academy of Science (2001), Disposition of high-level waste and spent nuclear fuel – the continuing social and technical challenge.

National Academy of Science (2003), ONE STEP AT A TIME, The staged development of geologic repositories for high-level radioactive waste.

Nuclear Waste Management Organization (NUMO) (2016) Numo no Taiwa-katsudo no Torikumi (NUMO's Activities of Public Dialogue), Document No. 2 of the 28th meeting of Radioactive Waste Working Group, METI, 1 September 2016. http://www.meti.go.jp/committee/sougouenergy/denryoku_gas/genshiryoku/houshasei_haikibutsu_wg/pdf/028_02_00.pdf.

Organisation for Economic Co-operation and Development, Nuclear Energy Agency (OECD/NEA) (1977), Objectives, concepts and strategies for the management of radioactive waste arising from nuclear power programmes.

Organisation for Economic Co-operation and Development, Nuclear Energy Agency (OECD/NEA) (1995), The environmental and ethical basis of geological disposal of long-lived radioactive waste.

Organisation for Economic Co-operation and Development, Nuclear Energy Agency (OECD/NEA) (2011), Reversibility and retrievability (R&R) for the deep disposal of high-level radioactive waste and spent fuel.

Saigo, T., S. Komatsuzaki and H. Horii (2010), Decisive factors of the dispute regarding high-level radioactive waste repository siting at Toyo-Cho, Kochi, Japan: An analysis of political process and possible solutions. *Shakai-Gijustu Kenkyu Ronbun-syu*, 7, 87–98. (in Japanese)

Science Council of Japan (2012), Kaito: Ko-reberu Hosya-sei Haikibutsu no Shobun nitsuite [Reply: On final disposal of high-level radioactive waste], September 2012. (in Japanese)

Sugawara, S. and K. Juraku (2010), Ko-level Hosya-sei Haiki-butsu Saisyu Shobun-jo no Ricchi Process wo meguru Kagaku-gijutsu-shakai-gaku-teki Kosatsu: Gen-patsu Ricchi Mondai kara-no 'Kyokun' to Seido-sekkei no 'Shippai' [A sociology of science and technology study on the siting process of high-level radioactive waste disposal plant: 'Lessons' from nuclear power siting and 'failure' of institutional design]. *Nenpo Kagaku-Gijustu-Shakai* (Japan Journal for Science, Technology and Society), 19, 25–51. (in Japanese)

Tashima, T., *Daremo Shiranakatta Chiisana-machi-no 'Genshiryoku Senso'*, Tokyo: Wac, 2008. (in Japanese)

8 Health effects of radiation exposure from the Fukushima Daiichi nuclear disaster in Japan

Discussing uncertainty in the context of environmental issues

Yuji Tateishi

1 Perspective

Six years have passed since the Fukushima Daiichi nuclear disaster of March 2011 produced widespread radioactive contamination. However, conflicts continue regarding the health effects of radiation exposure resulting from the accident. On one hand, many people express serious concerns about the possibility of radiation-related cancer and other diseases, maintaining that our scientific knowledge about the effects of radiation remains incomplete.[1] On the other hand, government representatives have adopted the position that sufficient research has been accumulated on radiation effects and that there is and will be no observable health effects of the exposure that has occurred due to this disaster. The antagonism between the two sides has gone far beyond friendly exchanges of views and has involved extensive attacks on the combatants' character and vested interests, to the point that reasoned discussion is now difficult (Nakamura 2013).

The extent to which uncertainty characterizes the current state of scientific research has been a key point of dispute between the two sides. The following quotations illustrate the typical position adopted by government-affiliated researchers on this issue:

> Regarding whether 'no effects are recognized' indicates that 'there are no effects', the reply would be that, scientifically, one cannot say that there are 'none'. When asked whether this means we 'don't know', the answer would be that such is the case. …
>
> Before effects 'are recognized', scientific debates occur between researchers who argue that they 'are not recognized' and those who say that they 'are recognized'. While this debate continues, if researchers on both sides directly release their contradictory findings or theories publicly, it will be difficult for the general public to know which findings are trustworthy. As long as scientific debate continues, scholars should consistently express to society the position that scientific uncertainty exists, so as not to cause confusion.
>
> (Nagataki 2012: 74)

Shigenobu Nagataki, a doctor who conducted research on the increase in cases of childhood thyroid cancer related to the Chernobyl nuclear accident (see Shibata *et al.* 2001), has played a central role in guiding the government's measures against radioactive contamination since the Fukushima accident. He was appointed as the joint chief investigator for the Chief Cabinet Secretary's Working Group on Risk Management of Low-dose Radiation Exposure (hereafter 'LRE Working Group'). Nagataki and many government-related researchers have adopted the viewpoint expressed in the above quotation, which acknowledges the existence of scientific uncertainty in general but does not concretely explain its specific content.

However, has this stance really been effective? This paper focuses on the possibility that the government-affiliated scientists' strategy of engaging in discussions without mentioning the uncertainty concretely has become an obstacle to meaningful dialogue with researchers and citizens who hold different perspectives. Japan has experienced a recent upsurge in nationwide anti-nuclear advocacy, triggered largely by public concerns about radiation exposure; many of the participants in these anti-nuclear protests have expressed their frustration at the government's stance regarding radioactive contamination (Oguma 2013). The fruitless nature of unresolved disputes over radiation risk could erode trust between the government and the public and could consequently pose a major obstacle to dialogue on Japan's future energy policy. With this issue in mind, the present paper analyses the responses of experts to and the application of scientific knowledge regarding radiation exposure.

2 Analytical framework

Since the late 1980s, when Beck (1986) articulated the concept of risk society, the idea of uncertainty has received widespread attention in sociology, although the interpretation of this word has varied considerably among scholars. Even in the context of environmental issues and health effects, the concept of uncertainty has various implications. Since the meaning of uncertainty tends to be rather indistinct without a reference point, the concept is sometimes defined in contrast to other similar concepts (i.e., Wynne 1992; Stirling 2007). Stirling argues that 'risk' is present when the outcomes and probability of an occurrence are clear, that 'ignorance' prevails when both are unclear, and that 'uncertainty' arises when the former is understood but the latter is unknown (the opposite combination is described as 'ambiguity'). In the case of low-dose radiation exposure, although the outcomes (i.e., health hazards such as cancer) are known to some extent, uncertainty over whether these outcomes will actually occur and, if so, on what scale has emerged as a primary focal point.

In environmental sociology and in science and technology studies, constructivist approaches have focused on the rhetorical use of uncertainty (Campbell 1985; Hannigan 1995; Adachi 2001). This group of studies excludes the question of whether uncertainty exists objectively, analysing instead the role of uncertainty within science and technology debates. These studies have demonstrated that in disputes regarding environmental problems, it is considered important to show how the opponent's argument is beset with uncertainties.

Other, more dynamic approaches have also been applied to research activities connected to claims of uncertainty. Such studies have examined, for example, the process of producing uncertainty around the reliability of DNA tests in criminal trials (Lynch 1998) and cases in which experts impose a specific interpretive frame without recognizing the uncertainty of their own claims and consequently damage people's trust in them (Wynne 1996). Levidow (2001) outlined the existence of a dynamism whereby precautionary considerations about environmental issues lead to the undertaking of broad studies, which in turn can result in the discovery of new uncertainties. These studies have established that uncertainty is not something that exists in advance or is simply asserted; rather, it is continually shaped by the actions of scientists and surrounding actors.

In light of the perspective described above, this paper will highlight the question of whether an instance of uncertainty being discussed is 'specific'. Today, when we formulate policy decisions regarding environmental problems, considering uncertainty is unavoidable. With regard to low-dose radiation exposure, many researchers on the government side recognize the existence of uncertainty. However, in this case, uncertainty is handled superficially and in general terms; these researchers are not confronting any uncertainties directly or specifically. Here, we should distinguish 'specific uncertainty' (vagueness or disagreement that is clearly expressed in relation to specific knowledge or research activities) from 'ambiguous and vague uncertainty' (unease). The concept of specific uncertainty was stimulated by Robert Merton's concept of 'specified ignorance', which is defined as 'the express recognition of what is not yet known but needs to be known in order to lay the foundation for still more knowledge' (Merton 1987: 7). Merton described the opposite of 'specified ignorance' as follows: 'The familiar kind of a general, rote, and vague admission of ultimate ignorance serves little direct cognitive purpose though it may have symbolic significance in reminding us of our limitations. … And vague questions evoke dusty answers' (Merton 1987: 7). Although Merton's main interest was in illustrating the social mechanisms of academic progress, I will apply his distinction to analyse the speech and behaviour of scientists dealing with environmental issues.

2.1 Materials for analysis

This paper compares the activity of two working groups, both created for the purpose of establishing standards regarding low-dose radiation exposure. As Kuchinskaya (2013) points out, the politics of formal representations – standards, thresholds, and visual mapping – is central in the social process of addressing radioactive contamination; subtle changes to environmental standards can significantly limit the scope of risks and make them invisible.

The first of the two groups to be examined is the aforementioned LRE Working Group, led by Goshi Hosono (Japan's nuclear disaster minister at the time), which met from November to December 2011 to assess the risks associated with low-dose radiation exposure. It was formed in the context of the government's shift from immediate post-disaster response activities to long-term policies, which included the reworking of 'no-go' zones. Following the working group's conclusions, the government ultimately adopted a 20 mSv (millisieverts) annual radiation exposure limit for policy purposes. This standard subsequently became the primary basis for measures related to radioactive contamination in Fukushima and nearby prefectures, although many citizens' groups criticized it because they believed that the standard forced residents to bear unacceptably high risks (Onai and Shirabe 2012).

The second working group, associated with the Food Safety Commission of Japan, met from April to July 2011 and was called the Working Group on Assessing the Food Health Impacts of Radioactive Materials (hereafter 'FSC Working Group'). The Food Safety Commission is a permanent administrative agency responsible for assessing food risks based on scientific knowledge. The FSC Working Group's purposes were to reassess the temporary standards regarding radioactive materials in food that had been imposed immediately following the nuclear disaster and to provide guidance for the transition to long-term measures (Hirata-Kimura 2013). Although these two working groups were established for similar reasons, we will see that they confronted uncertainty in completely different ways.

As materials for analysis, this paper utilizes the minutes of the meetings of both working groups, which have been published online.[2] It employs a qualitative analysis focusing primarily on the word 'uncertainty' (*fukakujitsu* in Japanese) and the content of statements related to uncertainty, as well as some simple quantitative analyses (e.g., counting the number of appearances of words related to uncertainty). It highlights the relationship between scientific knowledge and the social positions of the people who assert it (Martin 1988; Matsumoto 2002; Bocking 2004; Tateishi 2011).

3 The LRE Working Group: lacking discussions of 'uncertainty'

Table 8.1 displays the results of a search for the word 'uncertainty' throughout the full texts of both working groups' meeting minutes. The LRE Working Group minutes used the word just three times, compared to 50 times for the FSC Working Group Furthermore, the LRE Working Group also made few references to academic research, as shown by the infrequent use of the word 'article' (*ronbun* in Japanese).

Instead of going back to specific articles and considering their uncertainty individually, the LRE Working Group primarily discussed the application of the 'international consensus' to problems in Japan, as well as how to persuade the general public to accept it. Representative examples of international consensus

Table 8.1 Frequency of use of the words 'Uncertainty', 'Article', and 'Radiation' in working group minutes

	Uncertainty	*Article*	*Radiation**
LRE Working Group (8 Meetings)	3	47	544
FSC Working Group (9 Meetings)	50	209	305

Notes: * 'Radiation' included for comparative purposes.

regarding the effects of radiation include the reports of the United Nations Scientific Committee on the Effects of Atomic Radiation (UNSCEAR) and the recommendations of the International Commission on Radiological Protection (ICRP),[3] two agencies whose consensus positions are cited repeatedly in the LRE Working Group's minutes (UNSCEAR 2008, 2011; ICRP 2007). Following are some examples:

YASUHITO SASAKI: Thinking linearly is … reasonable. It's not that wrong. The ICRP discusses this on page 99 of the publication. (Seventh meeting)

KAZUHIKO MAEKAWA: [This article is] not discussed by a trustworthy international agency. It appears strange to us too. (Second meeting)

KAZUO SAKAI: This is numerically in accordance with the UNSCEAR, so I think its fine. (Eighth meeting)Such statements asserting that the content of a proposed report was acceptable because it followed UNSCEAR or rejecting an article because it was not discussed by a 'trustworthy international agency' were made innumerable times at the group's meetings. All members of the LRE Working Group were involved in the fields of radiation or nuclear energy (Table 8.2). They treated the 'international consensus', which they had participated in creating, as self-evident.

3.1 The uncertainty of the 'international consensus' and how the LRE Working Group overlooked it

Here, international consensus can be seen as a sort of 'discommunication' technique by which government-affiliated researchers block detailed discussion of uncertainty and prevent communication with the other side, thereby controlling the situation (Adachi 2001). Various criticisms have been levelled against this approach by the LRE Working Group (e.g., Onai and Shirabe 2012; Hayashi 2012). These criticisms focus on the inappropriateness of Nagataki's interpretation of the 'international consensus', as well as contending that this so-called consensus is actually characterized by diversity and uncertainty.[4]

The issue of childhood leukaemia after the Chernobyl disaster is one example of a grey zone relating to this supposed international consensus. The 2008 UNSCEAR report, which was published as UNSCEAR (2011), concludes that 'so far there are few studies available and little convincing evidence to suggest a measurable increase in the risk of leukaemia among those exposed as children to the radiation resulting from the accident at Chernobyl' (UNSCEAR 2011: D171). Although, at first glance, the report appears to deny any evidence of an increase in cases of childhood leukaemia, a close reading finds that it does cite a paper by A. G. Noshchenko *et al.* (2002) that claimed to find a measurable increase. That study reported that a Ukrainian childhood leukaemia group showed significantly higher estimated levels of exposure than the control group. However, since this study was said to have methodological problems and since another study by Davis *et al.* (2006), which also included Russian and Belarusian as well as Ukrainian

Table 8.2 Low-dose Radiation Exposure (LRE) Working Group Members and Their Fields of Expertise

Name	*Affiliation (At the Time); Career Background*	*Field of Expertise*
Keigo Endō	Kyoto College of Medical Science	Radiation Epidemiology
Kenji Kamiya	Fukushima Medical University, Hiroshima University	Radiobiology
Shunsuke Kondō	Head of Atomic Energy Commission, Formerly The University of Tokyo	Nuclear Power Engineering
Kazuo Sakai	National Institute for Quantum and Radiological Science and Technology, ICRP Committee 5 Member	Low-Dose Radiation Exposure, Radiobiology
Yasuhito Sasaki	Japan Radioisotope Association, Formerly National Institute of Radiological Sciences, Formerly ICRP Main Commission, Formerly UNSCEAR Chairperson	Radiology, Radiation Protection
Tomoyuki Takahashi	Kyoto University	Radiation Safety and Management Engineering
Shigenobu Nagataki (Joint Head)	Formerly Nagasaki University, Formerly Radiation Effects Research Foundation	Internal Medicine (Thyroid Cancer)
Ōtsura Niwa	ICRP Main Commission, Radiation Council Head, Formerly Tokyo University	Radiobiology
Kazuhiko Maekawa (Joint Head)	Head of the National Institutes for Quantum and Radiological Science and Technology's 'Radiation Emergency Medical Assistance Team', Formerly The University of Tokyo	Emergency Medicine, Radiation Emergency Medicine

Note: The above information has been gathered from various sources. The table has been created by the author.

children, found inconsistent results, the study by Noshchenko *et al.* was not accepted as part of the international consensus.

ICRP and UNSCEAR have repeatedly addressed the existence of this grey zone between certainty and uncertainty. For example, UNSCEAR (2008) lists existing radiation epidemiology research and systematically presents the strengths and limitations of each study in a large table. The articles underlying UNSCEAR's conclusions, including their limitations, are systematically noted. In addition, even articles that disagree with UNSCEAR's conclusions but have been published in international academic journals are included.[5]

The ICRP (2007) uses various approaches to create a consistent policy framework while relying on a substantial amount of scholarship, which includes expressions of uncertainty and disagreement. The ICRP's basic position is the linear no-threshold (LNT) model, which maintains that the exposure amount, expressed in

Sv (the effective dose), is proportionally related to an increase in deaths due to cancer. The process involved in calculating the effective dose was reviewed by the LRE Working Group. The following quotation comes from the minutes of the LRE Working Group's seventh meeting:

> ŌTSURA NIWA: The next related problem is that the speed of removal [of radioactive substances from the body] differs depending on the person (whether it is a child, etc.). Since these variations aren't included, there's uncertainty in the process of changing it to an actual Sv [value]. Then, the biggest [issue] is the use of the organ load coefficient. ... The types of cancer that occur in children and in fifty-year-olds are different, but they use an average of all of them. ... Therefore, in a bad case [the number] might be off several-fold.

Thus, there are some huge leaps in the process of calculating the effective dose, including adjustment for individual differences such as age, sex, and race and for environmental differences such as kinds of nuclides and the form of exposure. When the LNT model is connected to the effective dose, it becomes the basis for a 'regulatory science' framework that addresses demands for measures that will consistently protect people from radiation (Jasanoff 1990). Jasanoff compared academic research and regulatory science, describing the latter as based on evaluations by the government and industry, emphasizing relevance to policy rather than originality, and unwilling to wait for solid proof due to considerable temporal constraints. From the policymakers' perspective, modifying the method of calculating the effective dose in response to every new set of research findings is undesirable in terms of policy consistency and efficiency. Therefore, the radiation dose evaluation process is fixed 'by convention' and hence 'not subject to uncertainty' (ICRP 2007: 78). In other words, the ICRP's recommendations are the product of a process by which experts and policymakers control and stabilize uncertainty based on various adjustments and mechanisms.[6]

The statement by Niwa quoted above clearly discusses the uncertainty in the Sv calculation process, and it therefore could be viewed as touching on specific uncertainty. However, as noted previously, the minutes of all LRE Working Group meetings use the word 'uncertainty' only three times, including the above statement. Furthermore, Niwa made this statement in response to Minister Hosono's question regarding the extent to which the internal exposure risk can be determined. Not a single example is available in which a member of the LRE Working Group raised the topic of uncertainty independently.[7]

Therefore, in the LRE Working Group's discussions, research – even when in a grey zone – was placed within a binary framework, based on whether effects had been officially acknowledged by sources of international consensus. Although UNSCEAR's reports used nuanced expressions, the conclusions reached by this entity were presented to the working group in a very simplified manner – indicating, for example, that confirmed influence on residents after the Chernobyl disaster was limited to children's thyroid cancer only. Thus, the risk of childhood leukaemia was ignored, as it was grouped with diseases for which no health influences had been observed.

3.2 Articles outside the 'international consensus': highlighting uncertainty

If one wishes to discuss the uncertainty of science concretely, discussions of individual articles are indispensable. However, as shown in Table 8.1, the word 'article' was rarely mentioned in this working group, although more often than 'uncertainty'. Furthermore, 36 of the 47 uses of this word recorded in the LRE Working Group minutes occurred during two meetings attended by researchers with critical perspectives.[8] Articles were mentioned only 11 times at the remaining six meetings combined. What was the reason for this?

The LRE Working Group invited scholars critical of current policies to explain their positions. Shinzo Kimura, a Dokkyo Medical University radiation hygienist, participated in the second meeting, and Tatsuhiko Kodama, a University of Tokyo medical scientist, attended the fourth meeting. Both have received attention for their efforts to evaluate radiation doses and assist decontamination efforts in Fukushima after the disaster. Both have also criticized the government's understanding of the effects of radiation exposure.

At the working group meetings, problems in the articles highlighted by Kimura and Kodama were pointed out rather harshly. Following are two portions of an exchange between Kodama and working group members regarding the research finding that the incidence of bladder cancer has increased due to the effects of radioactive cesium from the Chernobyl accident:

NIWA: Epidemiology provides an indisputable answer regarding thyroid cancer. However, regarding bladder cancer, it has not provided a clear answer. … Regardless of whether it's pathology or something else, I'd like you to recognize that something cannot be considered reality if it has not actually happened in reality.

KODAMA: You mention epidemiological reality. … Basically, epidemiological studies can state a conclusion after years have passed and numerous parameters have been combined. However, when considering future predictions, especially at the stage when such things haven't come about, the direction [in which the research is heading] must be considered.

NAGATAKI: I'll say it again; Professor Niwa's question is whether bladder cancer is epidemiologically increasing. As Professor Maekawa just said, please answer only yes or no.

KODAMA: That question itself is a bit problematic. … I am talking about the kinds of concerns and the kinds of diseases that arise when viewed from the perspective of genome science, or science regarding signalling and the epigenome, which has been rapidly progressing since the beginning of the 21st century.

Although this exchange does not deviate from the rules of scientific discussion, the main issue here is that critical and close examination of scientific articles was not conducted systematically. Such scrutiny was directed more toward the research introduced by Kodama and Kimura. As we will see in the next section, the FSC Working Group's discussions identified various limitations in the type of scholarship that UNSCEAR and the ICRP have adopted as the basis for their consensus; however, no such examination critical, systematic examination of these studies took place in the LRE Working Group.

4 The FSC Working Group and specific uncertainty

When one is attempting to verify the validity of an alleged international consensus, it is indispensable to first refer to the individual articles that serve as the basis for this consensus, check their validity, and then conclude whether a consensus can indeed be derived from them. The FSC Working Group's discussions provide an example of this process.

The composition of this working group (Table 8.3) was in sharp contrast with that of the LRE Working Group. The original 13 expert members were specialists in such fields as medicine and food safety, the health effects of chemicals in the environment, and cancer epidemiology. However, virtually none of them specialized in research on the effects of radiation. Even though nine researchers working in radiation and related fields participated as expert witnesses, the group's own members actually led the discussions. This arrangement of membership and discussion style is typical of the FSC: discussion participants are usually not limited to experts on the chemical in question, and the regular expert members assume a leading role. On the issue of radiation exposure, this group composition made it easy for the members to maintain some distance from the existing radiation protection framework and reconsider the 'international consensus' through a bottom-up approach.

The bibliography provided at the end of this working group's proposed assessment is revealing. A three-level assessment was applied to the 51 listed studies on low-dose radiation: 'A: Consulted', 'B: Used as a reference when consulting other works', and 'C: Other'. In this way, articles that served as a basis for the report's conclusions were distinguished from those that were known to exist but were not used directly.

This outlook, which highlighted individual articles and reached conclusions based on them, was continuously emphasized throughout the working group's deliberations. Some of the expert witnesses, whose expertise was related to radiation risk argued that the assessment should be based on the reports and recommendations of the ICRP and similar international organizations.[9] However, this perspective was put on the back burner for all practical purposes, as Yasushi Yamazoe, the chairperson of the working group, stated at the group's seventh meeting that, although 'in the end they will be included, at first we have to evaluate individual articles'. In addition, though every member was provided with the ICRP's recommendations, the group's overall impression of them was not favourable. An expert member of the working group whose expertise is environmental toxicology stated, 'I couldn't understand what these Japanese [terms] mean' (Chiharu Tōyama, first meeting). The ICRP's reports have a unique nomenclature combining scientific facts and policy-oriented statements, as a result of their attempts

Table 8.3 Food Safety Commission (FSC) Working Group members and their fields of expertise

Name	*Affiliation (At the Time); Career Background*	*Field of Expertise*
Expert Members		
Ginji Endō	Osaka City University	Industrial Hygiene, Environmental Hygiene
Takashi Kawamura	Kyoto University	Preventative Medicine, Epidemiology
Hiroshi Satō (Chairperson Representative)	Tohoku University	Environmental Toxicology
Shoichiro Tsugane	National Cancer Centre	Cancer Epidemiology
Reiko Teshima	National Institute of Health Sciences	Food Safety, Immunotoxicity
Chiharu Tōyama	The University of Tokyo	Environmental Toxicology
Ken'ichi Hanaoka	National Fisheries University	Fisheries Biochemistry
Makoto Hayashi	National Institute of Health Sciences	Genotoxicity
Katsuyuki Murata	Akita University	Environmental Health Sciences, Epidemiology
Yasushi Yamazoe (Chairperson)	Tohoku University	Pharmacokinetics
Midori Yoshida	National Institute of Health Sciences	Toxic Pathology
Jun Yoshinaga	The University of Tokyo	Environmental Chemistry, Exposure Assessment
Hideki Wanibuchi	Osaka City University	Chemical Carcinogenesis, Experimental Pathology
Expert Witnesses		
Tomohiko Iwasaki	Tohoku University	Reactor Physics
Yasuhito Sasaki	Japan Radioisotope Association, Formerly the National Institute of Radiological Sciences, Formerly ICRP Main Commission Member, Formerly UNSCEAR Chairperson	Radiology, Radiation Protection
Hideo Sugiyama	Matsuyama University	Radiation Hygiene
Toshio Sofuni	Formerly National Institute of Health Sciences	Genotoxicity
Tomotaka Sobue	National Cancer Center Research Institute	Cancer Epidemiology
Yukio Takizawa	Akita University	Public Health, Environmental Health

Name	*Affiliation (At the Time); Career Background*	*Field of Expertise*
Tadao Terao	Formerly Food Safety Commission, Formerly National Institute of Health Sciences	Pharmacology
Keiichi Nakagawa	The University of Tokyo	Radiation Therapy
Junko Matsubara	Formerly Nuclear Safety Commission, Formerly Yokohama City University	Risk Science

Note: The above information has been gathered from various sources. The table has been created by the author.

(explained above in section 3.1) to establish consistent findings useful for policy purposes. Such terms were difficult to understand for group members whose speciality was outside the field of radiation, even for specialists in epidemiology or chemical risks. Furthermore, some ICRP reports are very long and quote other ICRP or UNSCEAR reports in a layered fashion. Participants repeatedly made remarks such as 'It's hard to get back to the original articles. Instead, I just found a statement that the numerical values should be about this much' (Yamazoe, fourth meeting). Thus, the FSC Working Group chose not to rely on the 'international consensus' and instead based its discussions directly on individual articles.

This working group's deliberations examined individual articles on the risk of low-dose radiation exposure and frequently pointed out the limitations and uncertainties of each one. Various articles were brought up for discussion, ranging from lifespan surveys of atomic bomb survivors to articles on astronauts exposed to cosmic radiation and studies on the effects of the Chernobyl disaster. The aforementioned study by Noshchenko *et al.* (2010) of Ukrainians impacted by the Chernobyl disaster was discussed in the following manner:

> Takashi Kawamura: [Dose estimation] was decided based on memories and interviews, though efforts were made to formalize interviews. The unavoidable recall bias that accompanies case control studies cannot be denied. There were probably originally not four but five categories. … It appears that they were combined because [with five categories] the number of cases would be reduced and insignificant.
>
> (ninth meeting)

In addition to issues of research bias, such as the fact that cancer patients will inevitably be more attentive to having been exposed to radiation, the arbitrariness of the category divisions in the article was pointed out. Such critical examination was conducted not only for this article but also for other studies.

Such technical discussions might seem far removed from societal decision making regarding the effects of radiation. However, this is not the case. In discussions about handling low-dose radiation exposure, the precautionary principle (which holds that countermeasures for environmental problems that could have substantial and irreversible effects should not be postponed due to uncertainty) is frequently raised. Environmental NGOs, which promote the application of this principle in all cases, and stakeholders of nuclear power plants and nuclear medicine, which advocate adherence to 'sound science', have long debated the application of precautionary actions (Nakagawa 2011; Hannigan 1995; Stirling 2007). In recent years, the government and its critics have not appeared to differ greatly in their recognition of the importance of precautionary responses. However, the differences become more evident when they move away from this general public stance and discuss specific uncertainty:

> TŌYAMA: This is very unclear and vague even in the [Noshchenko *et al.*] article; whereas on one hand, effects of comparatively low amounts, low radiation doses, have been reported, there are also articles that criticize this and then counterarguments against these critiques. Therefore, after all, we can't ignore this. (eighth meeting)

While discussing individual points, an epidemiologist member asserted that 'after all, we can't ignore this', but 'figures can't be derived' (Kawamura, ninth meeting). At another meeting, a group member asserted that 'collecting data [on bladder cancer] is very important, even if it cannot be included in benchmark values' (Hideki Wanibuchi, sixth meeting). In these discussions, one possible position is to exclude such scientifically uncertain questions as 'not formally acknowledged', whereas another position is to recognize the uncertainty and seek to have them reflected in policies in some form. Such differences in stances assume a concrete form when referring to specific uncertainty.

5 Negative self-reference

As discussed above, although both the LRE Working Group and the FSC Working Group ostensibly considered uncertainty, a detailed quantitative and qualitative analysis of the minutes of their meetings clarifies that the two groups treated the concept very differently. Compared to the latter group, the former barely addressed the degree of uncertainty present in research on low-dose radiation and took a strongly negative stance toward research inconsistent with the accepted international consensus.

In this way, the LRE Working Group's discussions were characterized by a lack of negative self-reference, which can be described as 'venturing to expose via admission the structural issues one bears' by 'going out on a limb to refer to the gap … between the form of knowledge one is seeking and the real state' (Matsumoto 2002: 160–161; see also Tateishi 2014). Wynne (1996) referred to such behaviour among scientists as 'critical self-examination' that aims to 'display the uncertainties in scientific knowledge, and at the same time expose as negotiable science's definition and role in relation to other social groups'. In contrast to the LRE Working Group, by referring to individual articles, the FSC Working Group clarified the uncertainty of this consensus, as well as of the articles upon which the consensus is based. The working group was also conscious of the uncertainties regarding the reliability of the research findings upon which the working group is relying. The LRE Working Group, meanwhile, made no attempt to apply to its own endorsement of the 'international consensus' the logic that it used to criticize studies that departed from this consensus.

5.1 What makes negative self-reference possible?

Why could the FSC Working Group make discussions accompanied by negative self-reference? As discussed above, the composition of the two working groups differed greatly, and the FSC Working Group's composition made it easier for this group to maintain a distance from the 'international consensus'.

It is also notable that the FSC Working Group conducted closed discussions including only researchers from universities and similar experts. In the discussion quoted in section 3.2, Nagataki and other members asked Kodama to answer 'yes' or 'no' regarding whether those living near Chernobyl had exhibited any increased rates of bladder cancer. Such a binary way of asking questions does not allow for ambiguity and is a kind of rhetoric rarely used in scientific discussions involving uncertainty; rather, it is frequently used at political meetings as people try to force the opposition into their own arena and compel them to make commitments (Adachi 2001). In fact, in the LRE Working Group, politicians, researchers, local government heads, and other stakeholders were visiting participants in almost every meeting. Furthermore, this group's meetings were broadcast live on the Internet. If a question like one posed to Kodama by the FSC Working Group were asked in full view of the public, the respondents would be preoccupied with making defensive statements to avoid criticism. As Collins and Evans (2002) noted, citizen participation can change the form of discussions and obstruct dialogues between experts. In the case of the FSC Working Group, it appears that negative self-reference was possible because the meetings were a closed space accessible only to scholars. I believe that this practice was adopted partly because of the FSC's institutional independence and deliberative tradition.[10]

In addition, if non-experts participate in meetings, it is not easy for them to discuss specific uncertainty. The non-experts who participated in the LRE Working Group repeatedly spoke about uncertainty. However, the manner in which they described it was totally different from that used by the radiation experts. For example, one National Diet member attending the LRE Working Group's second meeting stated, 'Not everything can be measured with science. When there are things that are not understood, we should be more modest about the achievements of science.'

As Merton (1987) pointed out, when uncertainty is mentioned in a vague or generalized manner, it has only symbolic significance in 'reminding us of our limitations'. Such counter-arguments, which do not touch upon individual academic articles or scientific claims, will simply be put aside as 'differences of opinion'. However, when the contentions contained in an individual article are cited, the response of those present will become all the more specific and specialized. In the latter case, a considerable amount of ability and advance preparation is required to continue responding without becoming engulfed in specialized knowledge. However, it is rather difficult for most non-specialists to develop such ability, which Collins and Evans (2002) called 'interactional expertise'.

5.2 The influence of negative self-references

Although negative self-references influenced the form of the discussions at the FSC Working Group meetings, the ultimate outcomes of this influence are not entirely clear. The conclusion of the FSC Working Group's proposed assessment was that it is 'difficult' to make 'judgments regarding less than 100 mSv [of radiation] accumulated *over a lifetime*' (emphasis added). While this position could result in stricter regulations than those suggested by the LRE Working Group's selection of '20 mSv *annually*' (emphasis added) as its threshold, it remains within the range of the ICRP's recommendations. Since the FSC Working Group did not focus on the international consensus, it undertook considerable independent effort to arrive at essentially the same conclusion.

The FSC Working Group's tentative assessment was described as being 'divorced from the real world' as it had not achieved what it had originally set out to do (Matsunaga 2011): it did not elaborate on the annual exposure limit required to avoid exceeding the lifetime accumulated exposure value, the number of mSv of internal radiation that could be deemed safe in food, or the number of Becquerel (i.e., units of radioactivity) to which each food item's benchmark value should be limited. In fact, at the sixth meeting of the FSC Working Group, Tōyama proposed that the Food Safety Council, being 'a government administrative agency', should provide recommended radiation dose amounts of individual foods 'as assessments that meet real-world demands'. However, the chairperson replied by stating simply that 'there are really no basic data' and by asking, 'Is it possible at present?' In the FSC Working Group's discussions, the relationship between the assessment of scientific data and recommended food contamination measures remained vague from beginning to end; uncertainty became a reason for not making any detailed recommendations.

When the Ministry of Health, Labour and Welfare (MHLW) revised its residual radiation benchmark value of food products in April 2012, the FSC's assessment was hardly used as a basis. After discussions held in the Pharmaceutical Affairs and Food Sanitation Council, the ministry decided to list first the guideline value of the Codex Commission, which formulates international standards for food, as the basis for its proposed maximum permissible annual radiation dose level of 1 mSv (MHLW 2012). Ironically, its policy judgments were ultimately made on the basis of the 'international consensus'. This overlooking of the FSC Working Group's efforts shows that the group's deliberative process was not fully aligned with the ideal type of negative self-reference, in which recognizing the contradiction between providing assessments directly connected to policy decisions and clarifying the limitations of those assessments is essential (Matsumoto 2002).

During the public comment period on the FSC Working Group's tentative assessment, more than 3,000 comments were received (*Tōkyō Shimbun* morning edition, 5 September 2011, p. 2; Hirata-Kimura 2013). Many commenters' interest focused on the group's unclear connection to policy. Arguing that a lifetime maximum exposure amount of 100 mSv should lead to a more restrictive annual benchmark value, they expressed the view that the selection of expert witnesses had been problematic and asked whether the assessment had been verified neutrally (for example, the National Parents Network to Protect Children from Radiation 2011; Hirata-Kimura 2013). Some citizen groups, including the National Parents Network, held lecture meetings presented by trustworthy experts so as to grasp the meaning of the FSC Working Group's tentative assessment before they formulated public comments. Thus, it can be observed that it was not easy for these citizens to keep up with the working group's deliberations on their own. Furthermore, the working group members did not actively share their understandings outside their meetings. Therefore, citizens' groups focused on those who participated in the meetings and on the conclusions expressed in the final reports, without referring to the discussion process. In such a situation, even if discussions regarding specific uncertainty are conducted, the scope of influence ends up being small.

6 Conclusion

This paper has examined the conditions that enable discussions on matters of specific uncertainty. According to the results, one important condition is that discussion participants must be located institutionally distant from an existing consensus. Negative self-reference, which systematically considers uncertainty (including uncertainty regarding the reliability of the research findings upon which one is relying), resulted in the elucidation of differences between the various members' positions and values. In addition, this study has shown that it may be difficult to achieve negative self-reference when non-experts participate in the discussion, although such participation is important to achieve meaningful impact not only within the expert community but also on society in general.

At present, many citizens are demanding greater participation opportunities in policy-making decisions regarding science and technology, and various attempts are being made to accommodate this demand (i.e., Matsumoto 2002; Bocking 2004). On one hand, the participation of diverse actors, who have different types of involvement with the concerned issue and who hold a wide range of political positions, might lead to instances in which experts are forced to address specific uncertainty (such as Niwa's statement cited in section 3.1). On the other hand, such broader participation might bring tension and antagonism into the discussions (Adachi 2001). From this point, we could go on to a further, more detailed examination of how a discursive space can be secured in such a way that the discussion does not deteriorate into a binary yes/no pair of options and instead encourages negative self-reference.

In doing so, we must note that non-experts often blend scientific views with daily life experience, historical experiences, and their personal values in developing their perspectives (i.e., the National Parents Network to Protect Children from Radiation 2011). Such non-experts' scientific views, which are connected with their social identities, often have 'a deeper and more complex consistency' than that recognized by a simplistic scientific model and are rich in insights and refinement when compared to those of experts (Wynne 1996). A differentiation between 'objective' uncertainty and subjective claims, especially if the former is viewed as having little relationship with citizens' participation, can deprive people of opportunities to speak about uncertainty (Wynne 1996, 2001).

In the current situation, we must also take into consideration an imbalance between the government and its opposition in terms of access to expert resources regarding radiation risks. In radiation-related fields, it has usually been difficult for citizen groups to develop and draw on a network of scientists who can provide a credible scientific foundation for questioning government actions, because of the strong connection between government officials promoting nuclear energy and scientists in these fields (Nakagawa 2011; Shimazono 2013; Tateishi 2013). Therefore, it is necessary to focus on structural problems in the production and distribution of scientific knowledge as a precondition for the convening of meaningful discussions that are capable of addressing specific uncertainty.

Acknowledgments

The original version of this paper was published in the *Japan Sociological Review* (*Shakaigaku Hyoron*), vol. 66, pp. 412–428. This paper was funded by Grants-in-Aid for Young Researchers (B) 26780296 and (B) 23730512.

Notes

1 For a typical example of citizens' concern over low-dose radiation exposure, see the National Parents Network to Protect Children from Radiation (2011). Soon after March 2011, many citizen groups emerged throughout Japan, the main purposes of which were to protect people, especially children, from radioactive contamination in food and water. In order to connect these groups with each other, the National Parents Network to Protect Children from Radiation (*Kodomo tachi wo hōshanō kara mamoru zenkoku nettowāku*) was established in July 2011.

2 I downloaded the LRE Working Group's meeting minutes from http://www.cas.go.jp/jp/genpatsujiko/info/news_111110.html and those for the FSC Working Group from http://www.fsc.go.jp/senmon/sonota/ (accessed August 03, 2014).

3 The UNSCEAR reviews and compiles academic articles on the effects of radiation exposure; based on these reviews, the ICRP presents principles and basic frameworks on protection from radiation exposure. From a history-of-science perspective, Nakagawa (2011) has critically discussed the formation of the ICRP's recommendations.

4 The assertions by the side criticizing the government were not merely criticisms of this 'international consensus'. Rather, some groups emphasized that the ICRP itself adopts an exposure limit of 1 mSv during normal times and demanded that the government should keep its 'promise' (i.e., National Parents Network to Protect Children from Radiation 2013).

5 Examples include an article by C. Busy, who heads an anti-nuclear power NGO, and another by M. Tondel *et al.* (2004), which noted an increase in cancer in Sweden after the Chernobyl disaster that far surpassed what one would expect from the ICRP model.

6 In the field of climate change, it has been pointed out that the 'stabilization' practice of experts and policy makers forms the background of the 'climate sensitivity' parameter, which has remained unchanged for a long time while models have been updated repeatedly (Sluijs *et al.* 1998).

7 The other two instances include one by a National Diet member who was specially participating in its second meeting and another by one of the ICRP main committee members who participated in its fifth meeting to provide explanations.

8 Yoshisada Shibata, who conducted epidemiological research on childhood thyroid cancer related to the Chernobyl accident along with Nagataki (Shibata *et al.* 2001), also participated in the second meeting to provide explanations. He highlighted problems of the academic papers on the Chernobyl disaster that are viewed as falling outside the 'international consensus'. This presentation contributed substantially to the high number of times that the word was mentioned during these meetings.

9 For example, 'Experts thoroughly have taken the time to engage in discussions at the ICRP and UNSCEAR and have concluded that such articles [research stating that even 10 mSv or less can have health effects] cannot be used, so to speak. If such conclusions were clearly shown and summarized, then a consensus could be arrived at to some extent' (Yukio Takizawa, seventh meeting).

10 Hirata-Kimura (2013) criticized the FSC's limited openness to the public, citing an excerpt from an interview with a staff member: 'Involving laypeople in the working group would have prevented it from making 'neutral and scientific' decisions.'

References

Adachi, S. (2001). Kōkyō jigyō wo meguru taiwa no mekanizumu: nagaragawa kakozeki mondai wo jirei toshite [The mechanism of dialogue over conflicts concerning public works project: The case of the Nagara River Estuary Barrage]. In Harutoshi Funabashi (ed.), *Kagai, higai to kaiketsu katei* [Environmental perpetration, suffering and the process of solving environmental problems] (pp.145–176). Tokyo: Yūhikaku.

Beck, U. (1986). *Risikogesellschaft auf dem Weg in eine andere Moderne.* Frankfurt: Suhrkamp.

Bocking, S. (2004). *Nature's experts: Science, politics and the environment.* New Brunswick: Rutgers University Press.

Campbell, B. L. (1985). Uncertainty as symbolic action in disputes among experts. *Social Studies of Science*, 15(3), 429–453.

Collins, H. M. and R. Evans. (2002). The third wave of science studies: Studies of expertise and experience. *Social Studies of Science*, 32(2), 235–296.

Davis, S.et al. (2006). Childhood leukaemia in Belarus, Russia, and Ukraine following the Chernobyl power station accident: Results from an international collaborative population-based case-control study. *International Journal of Epidemiology*, 35, 386–396.

Hannigan, J. A. (1995). *Environmental sociology: A social constructionist perspective*, London: Routledge.

Hayashi, M. (2012) Teisenryo hibaku mondai ha naze tsuzukunoka: fukko wo samatageru seifu no hoshasen anzenron [Why do the conflicts over low-dose radiation exposure continue?: the government's argument over radiation safety obstructs reconstruction]. Retrieved from http://archives.shiminkagaku.org/archives/2012/03/post-286.html (accessed 18 February 2017).

Hirata-Kimura, A. (2013). Standards as hybrid forum: Comparison of the post-Fukushima radiation standards by a consumer cooperative, the private sector, and the Japanese government. *International Journal of Sociology of Agriculture and Food*, 20, 11–29.

ICRP (International Commission on Radiological Protection). (2007). *The 2007 Recommendations of the International Commission on Radiological Protection.* ICRP publication, 103, Elsevier.

Jasanoff, S. (1990). *The fifth branch: Science advisers as policymakers*, Cambridge, MA: Harvard University Press.

Kuchinskaya, O. (2012). Twice invisible: Formal representations of radiation danger. *Social Studies of Science*, 43, 78–96.

Levidow, L. (2001). Precautionary uncertainty: Regulating GM crops in Europe. *Social Studies of Science*, 31(6), 842–874.

Lynch, M. (1998). The discursive production of uncertainty: The OJ Simpson 'Dream Team' and the sociology of knowledge machine. *Social Studies of Science*, 28(5–6), 829–868.

Martin, B. (1988). Analyzing the fluoridation controversy: Resources and structures. *Social Studies of Science*, 18, 331–363.

Matsumoto, M. (2002). *Chi no shippai to shakai: kagakugijutsu ha naze shakai nitotte mondai ka* [The failure of the science-technology-society interface]. Tokyo: Iwanami Shoten.

Matsunaga, W. (2011). Monkasho ga shokuhin anzen iinkai ni hitsuna shitsumonjo [The Ministry of Education submitted tough questions to the Food Safety Commission]. Retrieved from www.foocom.net/column/editor/4814/ (accessed February 18, 2017).

Merton, R. K. (1987). Three fragments from a sociologist's notebook: Establishing the phenomenon, specified ignorance, and strategic research materials. *Annual Review of Sociology*, 13, 1–28.

Ministry of Health, Labour and Welfare (Kōsei Rōdōshō) (2012). Shokuhin chuno hoshasei bussitsu ni kakaru kijunchi no settei ni kansuru Q&A nitsuite [Q & A about the setting of standards regarding radioactive materials in food]. Retrieved from www.mhlw.go.jp/shinsai_jouhou/dl/120412_2.pdf (accessed 18 February 2017).

Nagataki, S. (2012). *Genshiryoku saigai ni manabu hoshasen no kenko eikyo to sono taisaku* [Radiation health effects and countermeasures: lessons from past nuclear disasters]. Tokyo: Maruzen Shuppan.

Nakagawa, Y. (2011). *Hoshasen hibaku no rekishi: America genbaku kaihatsu kara fukushima genpatsu jiko made, Zohoban* [The history of radiation exposure: from the development of atomic bombs to the Fukushima nuclear disaster, an enlarged edition]. Tokyo: Akashi Shoten.

Nakamura, M. (ed.) (2013). *Posuto 311 no kagaku to seiji* [Science and politics after the disaster of March 11 in Japan]. Nakanishiya Shuppan.

National Parents Network to Protect Children from Radiation (Kodomo tachi wo hōshanō kara mamoru zenkoku nettowāku) (2011). Shokuhin anzen iinkai paburikku komento heno kenkai [Our opinion about public comments on the tentative assessment of the Food Safety Commission]. Retrieved from http://kodomozenkoku-news.blogspot.jp/2011/08/blog-post_24.html (accessed 18 February 2017).

National Parents Network to Protect Children from Radiation (2013). 1 miri siberuto kyanpen nitsuite 2 [On the campaign of 'one millisievert', part 2]. Retrieved from http://kodomozenkoku.com/1mSvcampaign/pg263.html (accessed 18 February 2017).

Noshchenko, A. G.*et al.* (2002). Radiation-induced leukemia risk among those aged 0–20 at the time of the Chernobyl accident: A case-control study in the Ukraine. *International Journal of Cancer*, 99(4), 609–618.

Noshchenko, A. G.*et al.* (2010). Radiation-induced leukemia among children aged 0–5 years at the time of the Chernobyl accident. *International Journal of Cancer*, 127, 412–426.

Oguma, E. (2013). *Genpatsu wo tomeru hitobito: 3.11 kara kanteimae made* [People who stopped nuclear power plants running: from March 11 to the meetings in front of the Official Residence of the Prime Minister]. Tokyo: Bungei Shunju.

Onai, T. and M. Shirabe (2012). Jumin dehanaku risuku wo kanri seyo: Teisenryo hibaku no risuku kanri nikansuru waking gurupu hokokusho ni hisomu sajutsu [Risk management of mind control?: Possible messages in the report by the working group on the risk management of low-dose exposures]. *Science journal Kagaku*, 82(3), 314–321.

Shibata, Y.*et al.* (2001). 15 years after Chernobyl: New evidence of thyroid cancer. *Lancet*, 358, 1965–1966.

Shimazono, S. (2013). *Tsukurareta hoshasen 'anzen' ron: kagaku ga michi wo humihazusu toki* [Fabricated arguments of radiation 'safety': when science takes the wrong way]. Tokyo: Kawade Shobo Shinsha.

Sluijs, J.van der*et al.* (1998). Anchoring devices in science for policy: The case of consensus around climate sensitivity. *Social Studies of Science*, 28(2), 291–323.

Stirling, A. (2007). Risk, precaution and science: Towards a more constructive policy debate. *EMBO Reports*, 8(4), 309–315.

Tateishi, Y. (2011). *Kankyo mondai no kagaku shakaigaku* [Environmental issues and the autonomy of science]. Sekai Shisosha.

Tateishi, Y. (2013). Hoshasen hibaku mondai niokeru hihanteki kagaku [Scientific criticism in the dispute over the risk of radiation exposure]. *Japan Journal for Science, Technology & Society*, 22, 31–46.

Tateishi, Y. (2014). Teisenryo hibaku womeguru tairitsu to fu no jikogenkyu [Disputes over the risk of low-dose radiation exposure and the critical importance of negative self-reference]. *Science journal Kagaku*, 84(5), 494–497.

Tondel, M.*et al.* (2004). Increase of regional total cancer incidence in north Sweden due to the Chernobyl accident? *Journal of Epidemiology and Community Health*, 58, 1011–1016.

UNSCEAR (United Nations Scientific Committee on the Effects of Atomic Radiation) (2008). *Effects of ionizing radiation: UNSCEAR 2006 report volume I.*

UNSCEAR (2011). *Sources and effects of ionizing radiation: UNSCEAR 2008 report volume II.*

Wynne, B. (1992). Uncertainty and environmental learning: Reconceiving science and policy in the preventive paradigm. *Global Environmental Change*, 2, 111–127.

Wynne, B. (1996). Misunderstood misunderstandings: Social identities and public uptake of science. In A. Irwin and B. Wynne (eds.), *Misunderstanding science? The public reconstruction of science and technology*, pp. 19–46. Cambridge: Cambridge University Press.

Wynne, B. (2001). Creating public alienation: Expert cultures of risk and ethics on GMOs. *Science as Culture*, 10, 445–481.

9 Nimby or civil disobedience? Conflicting framing and discourses of knowledge: wind turbine construction in Taiwan

Mei-Fang Fan

Controversies over large-scale renewable energy projects involving bitter disputes between private developers and local protestors have occurred in many countries (e.g. the United Kingdom, Australia, and the United States) (Bell *et al.* 2005; Cowell 2007; Groth and Vogt 2014). Local opposition to specific facility siting is often labelled as nimby (i.e. not in my back yard); this label is problematic because it wrongly characterizes local residents as irrational and selfishly unwilling to support projects benefiting the broader society (Burningham 2000; Cotton and Devine-Wright 2011). Although nimby suggests local opposition based on selfishness, it does not explain the reason that the individuals in question are opposed to the concerned development. Local opposition is partially rooted in the environmental justice implications of unevenly distributed risks and benefits as well as concerns over public health and pollution, and unhappiness with the decision-making process (Firestone 2011; Cotton and Devine-Wright 2011). Devine-Wright (2011) highlighted that nimbyism is a destructive, self-fulfilling way of thinking that risks undermining the fragile, qualified social consent that exists to increase renewable energy use. Breaking the cycle of nimbyism requires practising public engagement to better connect national policy-making to local areas directly affected by development projects.

Science and technology studies (STS) scholars view science as a strongly culturally embedded and dynamic social system, and have indicated that political and institutional forces produce structured gaps in knowledge, especially regarding chemical exposure and its effects on human health (Frickel *et al.* 2010; Hess 2010). 'Undone science' denotes areas of research that are left unfunded, incomplete, or generally ignored, but civil society organizations often identify as being worthy of further research. The concept of undone science has the potential to deepen the understanding of the systematic nonproduction of knowledge in the institutional matrix of state, industrial, and social movements (Frickel *et al.* 2010). Ottinger (2013) challenged the current notions of meaningful participation and informed consent and argued in favour of the STS-informed models of procedural justice, which should include ongoing opportunities for participation and consent as well as the obligation to continue to develop and refine knowledge relevant to residents' decisions.

In Taiwan, the development of renewable energy faces technical bottlenecks, higher costs than traditional energy, inadequate laws and regulations, and conflicts between local residents and owners. The Executive Yuan approved the One Thousand Sea and Land Wind Turbines programme in 2012 and established the Office Promoting One Thousand Sea and Land Wind Turbines Project, which is responsible for promoting and guiding wind power generation. The project adopted the strategies of From Land to Offshore, From Demonstration to Promotion, and From Shallow Sea to Deep Sea and developed Taiwan's wind farms progressively. By the end of July 2015, 28 wind farms had been developed. The Bureau of Energy, Ministry of Economic Affairs (2012), expects to complete the first offshore wind farm demonstration by 2020 and construct more than 1000 land–sea wind turbines by 2030, with a target of generating a total of 5,200-MW energy (Tseng 2015). The construction of wind turbines involves environmental and health risks and regulatory disputes as well as faces the opposition of local residents. According to statistics from the Ministry of Economic Affairs, on average, three of every ten wind turbines would be protested against. Some past protests include those against the wind power plants in Changhua Coastal Industrial Park, Sanzhi District, New Taipei City; Yuanli Township and Houlong Township, Miaoli County; and Xinfeng Township in Hsinchu County; the reasons for these protests included concerns regarding low-frequency noise and glare affecting driving safety; damage to windbreaks, crops, local ecology, and landscapes; and reduction in land value (Chen 2013).

These protests reflected the gap and decontextualization between the government-promoted green power policy and social cognition and values. Decisions and environmental impact assessments were limited by bureaucratic authoritarianism, which emphasized scientific positivism but failed to appropriately respond to disputes originating from society, values, ethics, and political operations. Faced with social controversy and opposition due to complex technological conflicts, a review of the framing dilemmas in policy discourse and process dynamics is required for developing regulatory policies.

Traditional policy analyses of environmental problems are based on the assumption that these problems are well-defined and the preferences and interests of the participants are externally fixed. By contrast, discourse analysis based on social constructivism acknowledges that interest regarding policy problems is developed during the policy-making process through the discursive practice of idea generation and exchange (Braun 1999). Discourse analysis focuses on the interaction discourse of policy actors, and explains how such a discourse is produced to compete with other actors (Dryzek 1997). It assumes that a certain discourse among policy actors can influence the way other actors perceive social reality and then change the meaning of preferences (Braun 1999). Therefore, discourse analysis is useful for understanding policy controversies (Schön and Rein 1994). This paper examines how meaning is constructed for policy problems through dynamic and interactive discursive practices during wind turbine construction and explores the framing of language and strategies used in the policy debate process of the wind turbine controversy in Yuanli Township, Miaoli County. After a brief discussion of the concept of risk governance and discourse analysis in policy studies, this paper analyses the differences in discourse and competing knowledge claims of relevant policy actors as well as the dynamics of policy disputes. In the concluding section, this paper presents a summary of the findings and policy implications.

Risk governance and discourse analysis

Risk governance includes a multifaceted and multiactor risk process as well as the consideration of contextual factors, such as institutional arrangements, political culture, and various perceptions of risk (Renn 2005). Risk governance in democratic societies necessarily works from the bottom up and is based on aggregating communal knowledge, experience, preferences, and concerns (Jasanoff 2010). Jasanoff (2010: 31–34) argues in favour of four focal points for the 'technologies of humanity' that can inform risk governance: framing, vulnerability, distribution, and deliberative learning. First, policy scholars emphasize that the quality of a solution to a perceived social problem depends on the adequacy of its framing. Few policy cultures have adopted systematic methods for asking questions about dominant risk framings. Second, traditional approaches to risk analysis view individual or social groups as passive objects, and their exposure is assessed through technical and expert calculation. Traditional approaches adopt risk-management judgment as the reference point for determining vulnerability, thereby neglecting the perceptions of the affected population and historical experience. Third, distributive effects and inequity are critical factors that evidently affect people's responses to harmful scenarios. Fourth, societies can actively reflect on ambiguity and assess the strengths and weaknesses of alternative explanations by constructing institutions of civic deliberation.

Regulatory practices and rule-making are shaped by social interactions and interdependencies between various actors, rather than being created through centrally controlled authority. Frames, framings, and discourses are of central importance in regulatory processes that reveal the construction of what is worth protecting, whom or what to protect, why, how, and by whom. Exploring how technological risks are governed requires examining how frames, actors, and knowledge intermingle with rule-making (Lidskog *et al.* 2010).

Hajer (1995) adjusted Foucault's concept of power and knowledge for understanding complex environmental policy controversy and stated that discourse analysis illuminates the social and cognitive bases of how policy problems are constructed (Hajer 1995: 15). The notion of 'storylines' is a central concept of Hajer's framework. Groups of actors being drawn to particular storylines as they reflect common interests results in the formation of 'discourses coalitions' (Hajer 1995: 12–13). Hajer defined discourses coalitions as 'the ensemble of 1) a set of storylines, 2) the actors who utter the story lines, and 3) the practices in which this discursive activity is based. Storylines are seen here as the discursive cement that keeps a discourse coalition together' (Hajer 1995: 65), thereby providing arguments for how these rationalities are upheld within policy discourse. For Hajer, discourse coalitions are linguistics based rather than interest based. Instead of traditional allegiances, the employed storylines and practices determine the coalition. Examples of opposing discourse coalitions can be seen in contemporary debates about renewable energy. New discourse coalitions have arisen through discursive strategies adopted in policy debates, and the traditional distinction of green groups has been refuted by the renewable energy debate (Stevenson 2009: 514).

Each storyline involves a different account of what is true and what counts as the representation of reality. Different discourse coalitions could draw on opposing knowledge claims as central elements in their distinctive storylines. Storylines play a critical role in positing actors, adding credence to the claims of certain groups and rendering those of other groups as less credible. Storylines also serve to create social and moral order within a given domain as devices through which actors are positioned and ideas of blame, responsibility, and urgency are attributed (Ockwell and Rydin 2006: 384). Benford and Snow (2000) proposed four framing processes: frame articulation, amplification, extension, and transformation. Frames are formed through discursive interactions and shift or evolve through the continual process of 'frame contesting' (Snow *et al.* 1986). Benford and Snow's (2000) framework of framing processes is useful for dealing with complicated cases where facts are uncertain, various policy disciplines are involved in defining and solving the problem, and high levels of professional, political, or popular conflict or conflict about appropriate policies and regulations are present.

Case study: wind turbine construction project in Yuanli Township

The proposal of the Qiwei Group and Tongwei Group (two subsidiaries of InfraVest, a Germany-based wind energy enterprise) to construct wind power plants in Zhunan Township, Tongxiao Township, and Yuanli Township in Miaoli County passed environmental impact assessment in 2007. Construction of the Yuanli wind turbines began in August 2012, approximately five years after the project passed assessment, and local residents were informed of the construction of the wind turbines in September 2012. Local residents who opposed the project established the Anti-Windmill Association to protest. The residents of Yuanli Township continued to take action 'against crazy windmills' (in Chinese, 'crazy' and 'wind' have the same pronunciation), which attracted the attention of the public and media.

The association gathered more than 200 residents to protest at the Bureau of Energy in January 2013, requesting that the Bureau of Energy revoke the licence. However, officials stressed that the project was filed in accordance with the Electricity Act and Environmental Impact Assessment Act and had gained the approval of the county government; thus, they had no grounds to revoke the licence. InfraVest began to construct the wind turbines on 19 February 2013, causing dissatisfaction among residents leading to their protest in the form of a hunger strike (Zhang 2012).

IntraVest had previously built wind turbines near the Yuangang Fish Harbour, and the Anti-Windmill Association queried whether this would damage the local environment and natural ecology. The association staged a sit-in protest to express their disagreement. The two sides clashed when the police asked the protestors to leave the site. A group of college students and local residents claimed that Intra-Vest had ignored the residents' opinions before constructing the wind turbines. The protestors tied themselves to the motor bases of the turbines with iron chains, and some even threw themselves beneath the construction vehicles to halt construction. Six residents and 14 college students were charged with coercion. InfraVest filed a lawsuit against seven members of the Anti-Windmill Association at a local court in June 2013, claiming civil tort and related damages valued at NT $10 million. InfraVest restarted construction on 26 July 2013, after conflicts with protestors. In 2016, the defendants were acquitted; however, the prosecutor appealed against the verdict. The Taichung Branch of the Taiwan High Court dismissed the appeal and acquitted all the protestors after a three-year trial. After two years of negotiation, the Anti-Windmill Association reached an agreement with the wind energy company to demolish the two controversial wind turbines and keep the other two. After their success at the lawsuit, the association continued to urge the government to draft new laws to protect the well-being of local residents (Ku 2014; Yang 2016).

The actors of the Yuanli case are grouped according to two opposing discourses, pro- and anti-windmill, each of which makes reference to knowledge claims in different ways. This case study conducted detailed discourse analysis to investigate the nature of conflicting knowledge claims as well as document analysis and in-depth interviews. This study also reviewed the wind power programme promoted by the Bureau of Energy, relevant policy white papers, official meeting records, environmental impact assessment reports, written information, official websites of owners, documents of civil society organizations, and news. Through purposive sampling, interviews were conducted with stakeholders of relevant policies to analyse the statements and knowledge of different stakeholders, including the Office Promoting One Thousand Sea and Land Wind Turbines Project, the Anti-Windmill Association, green groups concerned about the wind turbine issue, and experts from law and medicine fields.

Competing framing and knowledge-based claims of wind turbine construction

At least two contrasting discourses underlie the debate of the wind power project. That of the supporting side is 'economic and green energy development' which is considered essential for the transition to low carbon energy. Those who oppose construction are concerned with the negative impacts on health and ecology; thus, they prioritize regulations of safe distance. The gulf between these two sides is not easily reduced because the controversy involved scientific uncertainty and undone science.

Frame articulation

Each party has different views and discursive frameworks regarding construction of the Yuanli Township wind turbine. Wind power supporters stress the importance of renewable energy. In public meetings regarding wind power plant construction in Miaoli County, the Qiwei Group and Tongwei Group pointed out that local wind turbine construction would reduce the burden of local energy demands. InfraVest emphasized 'wind power is the best alternative energy', citing advantages including 1) wind turbines having the effects of wind resistance, sand fixation, and carbon reduction; 2) assistance for locals in implementing environmental maintenance; 3) increased employment opportunities to drive the development of local tourism and cooperate with related local activity planning; 4) promotion of local development and progress; and 5) training for local wind field guides and promotion of local tourism (InfraVest Wind Power Group 2013a).

The Anti-Windmill Association mainly stated the risks associated with wind turbines, and that low-frequency noise and dazzle produced during wind turbine operation were harmful to health. Anti-Windmill Association participants highlighted that residents of areas surrounding wind turbines in Taiwan (e.g. Houlong Township, Miaoli County and Shimen District, New Taipei City) had reported instances of tinnitus, palpitations, headaches, insomnia, frowsty bosom, and sleeplessness, and physical discomfort due to exposure to the long-term low-frequency noise of the wind turbines. Some Yuanli residents reported melancholia due to the running of the wind turbine and that the noise caused spiritual exhaustion and difficulty in falling asleep. The President of the Coastal Environment Development Association stated the following: 'I suffer from melancholia due to the wind turbine near my house, I wanted to commit suicide several times; fortunately, the situation improved after I changed my sleeping place, but I feel very inconvenienced' (Lai, 2013). Regarding 'wind turbine syndrome', an interviewed physician stated the following:

> The psychological and emotional impact will be substantial because the noise will affect hearing and will have a cumulative effect if severely noisy in a public place. The psychological impact is converted into physiological impact for most people, leading to mental and physical disorders. Such disorders affect brain and nervous system function and one's thinking ability. Such problems can cause melancholia or bipolar disorder, in turn potentially leading to insomnia. In serious cases, suicidal tendencies can occur, with some individuals resorting to violent behaviour. Regarding physical disorders, poor sleep negatively affects blood pressure, and high blood pressure can cause cardiovascular problems.

Wind turbine construction could affect life stability, quality of life, and the right to exist for residents in nearby areas. The construction affects not only human health but also livestock, fish, and bird ecology. Residents argued that wind turbine construction will make Yuanli a place where even pigs or dogs cannot live peacefully (Wild at Heart Legal Defense Association 2013b). Some Yuanli fishermen stated that local sea eels with high economic value and other offshore creatures might migrate or die because of low-frequency wind turbine noise sustained over long periods because sea eels have a natural nervous disposition and are sensitive to noise, water temperature, and strong light. Such an ecological change could negatively influence fishing ports and the livelihood of fishermen.

In addition to concerns regarding health and ecology, the risk perception of local residents about the wind turbine construction is affected by existing property values, social interaction, and cultural identity. A local coastal chief reported the presence of a cemetery within 30m of the turbine. The turbine could thus damage the feng shui (Qi 2012). Surveyed residents stated that wind turbine construction would impact local cultural beliefs:

> The first phase in Yuanli includes six turbines – four have been erected and two are situated in Hai'an and Hispin Villages. There are tombs only 20m away from a wind turbine. The protests and disputes over these two turbines are highly controversial; residents cannot bear the running of the wind turbine there because it is our tradition to respect our ancestors. Will our ancestors rest in peace if the wind turbine is installed there? They should respect our Chinese culture.
>
> (Interview with Anti-Windmill Association R01)

Amplification: competing interpretations of wind turbine noise

For the wind power operators, no accurate data was available to prove that wind turbine noise could negatively influence human health and environmental safety. The operator cited an overseas study showing that wind turbines can coexist with birds and stated that the development plan was in accordance with the legal procedures of the Law of Environmental Impact Assessment. The operator also held public meetings with the local people (InfraVest Wind Power Group 2013b).

InfraVest cited a study conducted in Sweden to criticize the so-called wind turbine syndrome, disregarding it as 'nonsense'. They also questioned the report regarding low-frequency noise and the 2km distance limit put forward by a professor from the New Zealand Anti-Windmill Association, highlighting that in a disputed case in 2013, the environmental court of New Zealand judged that the study was insufficient in its support of the 2km distance limit claim. The operators stressed the lack of empirical evidence suggesting that wind turbine noise has an adverse effect on the health of nearby residents and that no country regarded turbines as special noise facilities requiring regulations on distance. Therefore, they failed to return specifications for noise control standards (InfraVest Wind Power Group, 2013b).

The Anti-Windmill Association indicated that the wind turbines built by IntraVest were only 60–250m from the neighbourhood, which is far closer than the designated international safe distance. In the environmental assessment report of Zhunan regarding the Tongxiao and Yuanli wind power development projects, passed in 2007, InfraVest stated the following: 'The establishment of wind turbines shall be based on the principle of having great distance from building and regarding (wind turbines in the east–west direction) the distance between erections is approximately 210 to 270m' (three times the diameter of the blade), in principle, and that the distance between erections (wind turbines in the south–north direction) is approximately 350–450m to avoid chain accidents of other wind turbines or damage to the safety of nearby residents when a wind turbine has an accident that produces an explosion or collapse. Some residents stated that the operator had violated the commitment made in the environmental assessment report by making the spaces between several wind turbines too small and failing to position the turbines at the required distance away from dwellings. The residents claimed that the low-frequency noise produced by the turbine would interrupt the peace of local residents and threaten public safety. During an interview, the environmental group stressed that the government's safety distance regulations were vital for wind turbine construction. They stated that what we need is to have a safe distance and not to be against green power (Wild at Heart Legal Defense Association 2013b):

> We are against neither green energy nor wind turbines, but there should be specifications. If the safety specifications are reasonable and the government is willing to protect civilians, there is no problem with erecting 100 wind turbines in Yuanli, but if failing to conform to the specifications, wind turbines should not be allowed. At present, residents in Yuanli have a clear resolution, namely the need for a 'decree'. The current protest activity is a demand for a clear decree. Besides the unwillingness to set it up in my house, two accidents occur. If specifications are set in accordance with the specifications of international trends, there will be no problem with constructing the wind turbine, which should be constructed by the government rather than ordinary people.
>
> (Interview with Environmental Protection Group E01)

The Anti-Windmill Association applied to the court for a provisional injunction prohibiting construction of the wind turbine. However, the local Miaoli court determined that the Anti-Windmill Association did not have accurate data to prove that wind turbines affect living peace and health, that each person's perception of low-frequency noise is different, and that the noise can be dampened by installing sound insulation windows. The judges determined that InfraVest had legally acquired a licence for wind turbine planning and construction and that halting construction would cause greater damage to the company. Thus, the judges stated that because there were no significant damages, establishing a temporary state provisional injunction was unnecessary and the Anti-Windmill Association's protest application was rejected. Regarding the risk sanction, the court supported the risk maker on the basis of scientific evidence and cost benefits (Qi 2013). Some of the surveyed experts stated that the risk sanction could not be overlooked on the basis of the lack of evidence or insufficient evidence, whereas others stated that the noise would disappear and would not harm the human body:

> Why does the country give uncertain proof to its citizens? The government should address the current situation, rather than evaluating it according to previous experiences.

> Noise passes and disappears, but it cannot be said to have no impact. It is obvious that many people feel offended on seeing the wind turbines. Such the adverse effect of noise on one's mood could be viewed as a disease, which can affect a person's heart and blood vessels; psychiatrists can confirm this.
>
> (Interview with doctor D01)

Some local residents asserted that the wind turbine constructed by the manufacturer was too close to the residential area and questioned the environmental impact assessment procedures for the project, the imperfect governance regulations, and the operator's lack of adequate communication and consultation with residents. Members of the Anti-Windmill Association reframed their messages based on the idea of 'procedural justice' in decision-making, thereby criticizing the deficiency of the existing environmental assessment mechanism and legal regulations. According to the Law of Environmental Impact Assessment, only factors affecting the environment can become objects of environmental impact assessment, whereas nonenvironmental damage caused by private wind power construction, such as noise, is the subject of limited consideration. The current environmental impact assessment mechanism is deficient in various ways and thus fails to cover the assessment of vulnerability and justice. The public meeting held by the developer was only intended to spread unidirectional messages regarding wind turbine development and therefore could not clarify the knowledge and ideas of residents regarding the influence on culture and health. The Anti-Windmill Association stated the following in protest:

> We informed the Bureau of Energy that there had been no public participation in the turbine siting decision. We had no idea that such a large wind turbine was to be constructed or that problems existed in the procedures and administration of the project. The key issue is between the environment and ministries, and between the entire coast and humans. Just as the issue of migratory birds becomes apparent only when people pay attention to it and everyone works together, we are not called the Anti-Windmill Association as a protest, but because only in this way can we effectively spread the word about the issue of green energy in Taiwan.

A consulting engineer from the Anti-Windmill Association highlighted that according to data from the World Health Organization, long-term exposure to low-frequency noise of 32dB would influence health. The Anti-Windmill Association originally claimed that the distance between wind turbines and buildings should be classified into three levels and that public facilities and residences should be distinguished into two classes. The first class argued that wind turbines should not be set up within the scope of 'Mandatory regulation', while the second class agreed that constructing a wind turbine was acceptable provided the local residents 'agreed' and operators offered appropriate feedback. The Wild at Heart Legal Defense Association indicated that the noise control standard of Taiwan was 5dB and should have been reduced (Wild at Heart Legal Defense Association 2013a).

Regarding residents' questions regarding low-frequency noise and safe distance, the Bureau of Energy held a consultative conference on 20 August 2013 to map out a preliminary safe distance plan. They agreed on the following: 1) The setting distance of a wind turbine should be 1.1 times the height of the turbine in places without specific objections, such as distance from a road; 110m is the maximum required distance for turbines lower than 110m in height. 2) For construction of a wind turbine within a certain distance from buildings, consent should be obtained from local residents, and environmental impact assessments within 250m of said buildings should be in accordance with the provisions. This draft planned by the Bureau of Energy caused dissatisfaction from related industry operators, who thought that the regulation of 250 m was too harsh and would result in the cancellation of 70%–80% of planned wind turbines, which would be adverse to industry development. Moreover, the already constructed 314 wind turbines had caused no casualties and nuclear and thermal power plants, factories, and other facilities more dangerous than wind turbines had no formulated specifications for safe distances (Lai 2013).

The Wild at Heart Legal Defense Association was dissatisfied with the new scheme for safe distance specified by the Bureau of Energy. They thought that the information disclosure was incomplete, did not understand where 250m had come from, and questioned the measurement basis of noise control, which they thought reflected residents' mistrust of government regulations:

> Noise measurement is divided into low-frequency and super-low-frequency. Inaudible sound is harmful to humans, so how is this measured? We know that the act is nonsense, and we have protested our exclusion from participating in the formulation process. We subsequently exerted great pressure on the operators, but the wind turbine was constructed. Our premise is that the site of a wind turbine should be a standard distance from dwellings. This could solve parts of the problem because not everyone recognises low-frequency noise; however, the vibration of low-frequency noise affects our arteries, so not hearing it does not represent no harm; such noise could confuse the body about the difference between day and night.

Frame extension and reorganization of definitions: 'win–win green energy security'

The Anti-Windmill Association stressed not fighting against green power and demanded win–win green energy security. Problems in control regulations and procedures also highlight the differences between current wind turbine policy promotion and the cognition of civil society regarding green energy. The protest experience regarding foreign wind turbines showed that local people believe that most benefits are enjoyed by the developers, although operators highlight that such developments can promote community employment as well as subsidies and feedback from electricity prices. The training of relevant technical personnel and maintenance workers for wind turbines must rely on external support; thus, locals remain in a weak position. Environmental protection groups think that the government's promotion of renewable energy should emphasize small electric power stations distributed among the community. Such groups highlight foreign practices showing that community residents not only enjoy ownership of but also participation in the planning process of wind turbine construction. At present, many issues such as the most suitable location for construction are not discussed before implementation; therefore, it is necessary to provide public incentives for participating in the development of local wind energy, thereby promoting a move towards citizen-owned wind farms (Interview with Commissioner C01).

The Office of One Thousand Sea and Land Wind Turbines of the Bureau of Energy emphasized the necessity for 'communication', expecting the Anti-Windmill Association to clarify their doubts and a platform for coordination on the basis of the principles of being transparent, open, fair, and peaceful. The health risks of wind turbines still involve the problem of trust. The Office of One Thousand Sea and Land Wind Turbines thought that research funding was not a problem. The key questions were as follows: who would implement the project, who is acceptable to implement it, what are the procedures, and what degree of implementation could be accepted? (Interviewed officer W1).

The Bureau of Energy had promised to hold a hearing, and the Anti-Windmill Association claimed that public participation levels must be increased, information must be disclosed to achieve full and open discussion, relevant reports must be opened early for the disclosure of relevant information such as how the data put forward in the report would be calculated, and the name of the expert making the proposal. During an interview, the legal experts stated that each government department must strengthen horizontal connections and coordination and encourage each party involved in the decision-making process to adopt open and fair procedures to strengthen communication and build a relationship of trust. Energy should be understood in terms of its source, and humans and public health should not be ignored (Interview with Expert E02).

Local residents expressed demands and value propositions in various forms and through various actions on different occasions. Members of the Wild at Heart Legal Defense Association attempted to restrict wind turbine construction in a peaceful and nonviolent manner, but have been in conflict with the police on several occasions. At one instance, two members displayed protest banners to stop engineering vehicles on the road leading to a construction site and were sued by prosecutors. However, they were acquitted after a trial in Miaoli District Court; thus, the protest carried legitimacy. InfraVest has filed 46 civil and criminal actions against members of the Anti-Windmill Association. Although they ultimately agreed to withdraw the lawsuits, the protestors insisted on procedural justice. After the Miaoli District Court acquitted the protestors of all offences, the prosecutor appealed to the Taiwan High Court, which gave a final verdict of acquittal. During the trials, duty lawyers, students, and residents all advocated for safeguarding their hometown and the freedom of speech, considering themselves as representatives of housing justice to broaden their support base. One defendant, who was a resident, stated 'This was a justice delayed, but evil cannot prevail over good after all. We appreciate the judicial system proving our innocence'. In addition, one demonstrator, a village chief, stated 'After the demonstration, IntraVest demolished the wind turbine by themselves, indicating that they know they were wrong. We knew that justice was on our side and we would be acquitted eventually'. The Anti-Windmill Association asserted that although renewable energy development is an essential part of the energy structure of Taiwan, it must not be violently achieved (Ku 2014).

For local actors and opposing parties, local activism represents civil disobedience. In the studied case, although residents expressed opposition to the wind turbine construction policy through systematic channels, they did not obtain a comprehensive response from the authorities, and thus, they violated the current law regarding open and nonviolent means of communication with the authorities and public, and sought external claims to support their idea of safe distance.

The Anti-Windmill Association stated that they welcome the government promoting alternative clean sources of energy, but that it should develop a mechanism for local residents to participate in projects, rather than supporting large enterprises (Ku 2014; Yang 2016). Moreover, the Anti-Windmill Association continued to promote the Yuanli Grassroots People's Forum at the protest site, and local residents freely discussed local public issues and the imagined 'energy democracy'. Platforms such as the Grassroots People's Forum help to strengthen local communication, enhance citizens' identity and consciousness, and attempt to link wider communities and public interests.

Conclusion

This study indicated that discourse and risk perceptions differ among policy actors. In the studied case, the wind energy firm responsible for constructing the wind turbines stressed the importance of renewable energy and the benefits of the development project, whereas local residents and activists were concerned with the proximity of the turbines to their homes, worrying that the low-frequency noise produced by the machines may negatively affect their health and quality of life. The wind energy firm and opposition coalition cited scientific reports as their argument bases and refuted each other's claims.

Local opposition to wind turbine construction projects evinces the gap between green energy policies and local perceptions and social values as well as competing knowledge claims among policy actors. The promotion and construction of wind power generators in Taiwan represent utilitarian policy-based thinking, which neglects consideration of overall land usage planning and a sound legal system. In a situation where economic benefits are emphasized, the assessment of the influences of a sustainable environment, society, and culture are insufficient. Based on procedural justice, local Yuanli residents and the Anti-Windmill Association question the legitimacy of construction plans, defects of environmental impact assessments, and deficiency of current control specifications. Risk assessment of the influences of wind turbine location on health and ecology is mainly conducted with surveys of consultant firms and analyses of foreign professional studies as crucial bases. The formulation process of policy regulations fails to further clarify relevant scientific controversies and lacks sufficient dialogue and a communication mechanism between the related parties. The case of the wind turbine construction project described here reflects the problems of insufficient information and undone science as well as failure in implementing the informed consent principle.

This case shows that bureaucratic organizations formulate policies based on scientific research that although put forward by experts cannot obtain authority or legitimacy in knowledge, and only cause people to distrust the government and engenders conflicts between operators and community residents. The effectiveness of expert government knowledge relies on the knowledge-based propaganda of industry operators and is thus questionable. The deliberative citizen participation mode can provide an equally rational communication platform and action field for dialogue, encourage relevant policy actors to interact with one another, and negotiate and shape knowledge (Meadowcroft 2004). In addition to strengthening cross-sectoral coordination and clarifying policy goals regarding social welfare related to renewable energy, relevant responsible control organizations should be devoted to system reform to integrate citizens into policy development, construct participatory technology impact assessment mechanisms, and strengthen civil society by encouraging residents with experience in the local community to participate in the scientific knowledge development for the location and operation of wind turbines as well as the formulation of control regulations.

Decisions regarding the distance between wind turbines and buildings must be made in accordance with the agreement of residents within such distance, detailed design including disclosure and providing sufficient information, and consent obtained through open and transparent procedures. If a wind turbine construction plan obtains the agreement of nearby residents but scientific research shows an evident threat to public health and the environment or accidents occur, comprehensive negotiation and compensation mechanisms should be applied, where residents can either revoke their agreement or exercise their consent right again. In the operation phase after a wind turbine has been constructed, industry operators should establish a good neighbour partnership, devoted to learning more about the local environment, culture, and lifestyles and seeking solution-based schemes for the problems of sustainability for community organizations and residents to encourage local residents and members of civil society to participate in the supervision of safety maintenance, long-term environmental monitoring (e.g., influence on eel larvae and birds), and the strengthening of accountability mechanisms.

References

Bell, D., Gray, T. and Haggett, C. (2005). The 'social gap' in wind farm siting decision: Explanations and policy responses. *Environmental Politics*, 14(4), 460–477.

Benford, R. and Snow, D. (2000). Framing process and social movements: An overview and assessment. *Annual Review of Sociology*, 26(6), 611–639.

Braun, D. (1999). Interests or ideas? An overview of ideational concepts in the public policy research. In D. Braun and A. Busch (eds.) *Public policy and political ideas* (pp. 11–29). Northampton, MA: Edward Elgar.

Bureau of Energy, Ministry of Economic Affairs. (2012). *White Paper on Energy Industry Technology*. Taipei: Bureau of Energy, Ministry of Economic Affairs.

Burningham, K. (2000). Using the language of NIMBY: A topic for research not an activity for researchers. *Local Environment*, 5(1), 55–67.

Chen, Z. P. (2013). A variety of reasons for local opposition to wind turbines. *United Evening News*. page A4. Retrieved from http://www.coolloud.org.tw/node/74373.

Cotton, M. and Devine-Wright, P. (2011). NIMBYism and community consultation in electricity transmission networking planning. In P. Devine-Wright (ed.), *Renewable energy and the public: From NIMBY to participation* (pp. 115–128). London: Earthscan.

Cowell, R. (2007). Wind power and 'the planning problem': The experience of Wales. *European Environment*, 17(5), 291–306.

Devine-Wright, P. (2011). Public engagement with large-scale renewable energy technologies: Breaking the cycle of NIMBYism. *Climate Change*, 2, 19–26.

Dryzek, J. (1997). *The politics of the earth environmental discourse*. New York: Oxford University Press.

Firestone, J. (2011). Perceptions and preferences regarding offshore wind power in the United States: The leading edge of a new energy source for the Americas. In P. Devine-Wright (ed.), *Renewable energy and the public: From NIMBY to participation* (pp. 233–244). London: Earthscan.

Frickel, S., Gibbon, S., Howard, J., Kempner, J., Ottinger, G., & Hess, D. J. (2010). Undone science: Charting social movement and civil society challenges to research agenda setting. *Science, Technology & Human Values*, 35(4), 444–473.

Groth, T. and Vogt, C. (2014). Residents' precisions of wind turbines: An analysis of two townships in Michigan. *Energy Policy*, 65, 251–260.

Hajer, M. A. (1995). *The politics of environmental discourse: Ecological modernization and the policy process*. New York: Oxford University Press.

Hess, D. (2010). Environmental reform organizations and undone science in the United States: Exploring the environmental, health, and safety implications of nanotechnology, *Science as Culture*, 19(2), 181–214

InfraVest Wind Power Group. (2013a). The benefits of wind energy. Retrieved from http://infravest-twonline.com/infravest/TC/T06-7_benefit.pdf.

InfraVest Wind Power Group. (2013b). Is it necessary to make particular rules on distance of a wind turbine based on noise? Retrieved from http://infravest-twonline.com/infravest/TC/4-distance.htm.

Jasanoff, S. (2010). Beyond calculation: A democratic response to risk. In A. Lakoff (Ed.), *Disaster and the politics of intervention* (pp. 14–40). New York, NY: Columbia University Press.

Ku, H. (2014). Wind turbine protesters stage sit-in. *Taipei Times*. 17 April. Retrieved from http://www.taipeitimes.com/News/biz/archives/2014/04/17/2003588187

Lai, P. Y. (2013). Discussion on the safe distance regulations of wind turbines: critique of the hearing held by the Bureau of Energy. Taiwan Environmental Information Center. Retrieved from http://e-info.org.tw/node/88820.

Lidskog, R., Soneryd, L., and Uggla, Y. (eds.). (2010). *Transboundary risk governance*. London: Routledge Press.

Liu, L. R. and Wei, Y. J. (2013). Low frequency noise of wind turbines: public concern about harm to health. *Liberty Time Net.* Retrieved from http://www.libertytimes.com.tw/2013/new/aug/25/today-life1-2.htm.

Meadowcroft, J. (2004). Deliberative democracy. In R. F. Durant, D. J. Fiorino, and R. O'Leary (Eds.), *Environmental governance reconsidered: Challenge, choices, and opportunities* (pp. 183–217). Boston, MA: MIT Press.

Ockwell, D. and Rydin, Y. (2006). Conflicting discourses of knowledge: Understanding the policy adoption of pro-burning knowledge claims in Cape York Peninsula, Australia. *Environmental Politics*, 15(3), 379–398.

Ottinger, G. (2013). Changing knowledge, local knowledge, and knowledge gaps: STS insights into procedural justice. *Science, Technology & Human Values*, 38(2), 250–270.

Qi, R.-Y. (2012, November 22). Residents in Yuanli Township oppose to wind turbines. *United Daily News*, page B1.

Qi, R.-Y. (2013, June 20). Anti-windmill association requesting stop the construction of wind turbines: the District Court rejected the request. *United Daily News*, page B2.

Qiwei Group and Tongwei Group. (2011). Report on analysis on environmental influence difference of wind power plant construction project in Zhunan Township, Miaoli County, Tongxiao Township, Yuanli Township. Miaoli County: Qiwei Group and Tongwei Group.

Renn, O. (2005). *Risk governance: Towards an integrative approach.* Geneva: International Risk Governance Council.

Schön, D. and Rein, M. (1994). *Frame reflection: Toward the resolution of intractable policy controversies.* New York: Basic Books.

Snow, D., Rochford, E. B., Worden, S. K. and Benford, R. D. (1986). Frame alignment processes, micromobilization, and movement participation. *American Sociology Review*, 51(4), 683–696.

Stevenson, R. (2009). Discourse, power, and energy conflicts: Understanding Welsh renewable energy planning policy. *Environment and Planning C: Government and Policy*, 27, 512–526.

Tseng, G. R. (2015). Promoting one thousand sea and land wind turbines project. Energy Report, Retrieved from http://energymonthly.tier.org.tw/outdatecontent.asp?ReportIssue=201511&Page=16.

Wild at Heart Legal Defense Association, Taiwan (2013a). Guarding green energy: asking the government to regulate wind turbines siting and undertake health impact assessment. Retrieved from http://zh.wildatheart.org.tw/story/10/7333.

Wild at Heart Legal Defense Association, Taiwan (2013b). Inappropriate wind turbines: even pigs or dogs cannot live peacefully. Retrieved from http://zh.wildatheart.org.tw/story/10/7344.

Wu, Y.-P. (2011). The study on the wind power energy related environmental disputes and litigations. Unpublished MA thesis. National Tsing Hua University, Hsinchu.

Yang, O. (2016). Taiwan anti-wind turbine protesters acquitted after three-year trial. The News Lens. March 7. Retrieved from https://international.thenewslens.com/article/37599.

Zhang, X. T. (2012). Wind turbines go to Yuanli: residents protest against the project. *Liberty Time Net.* Retrieved from http://www.libertytimes.com.tw/2012/new/oct/20/today-north6.htm.

Conclusion

Toward a participatory and democratic risk governance

Ying-Feng Chen and Chien-Ming Hsu

Risk, energy and governance in East Asia: patronizing developmental states

At the beginning of this book, Hasegawa illustrated how the Fukushima nuclear accident was a structural disaster highly associated with system failure. First of all, Japan's energy insecurity and national defence demands have made nuclear power a favourable option for the government and citizens. With the government being the main agent promoting nuclear plants, safety regulations were therefore compromised by political expediency. The verdicts in court also left very little room for the general public or local community to demand greater government scrutiny via nuclear regulation. Furthermore, Japan's closed political system also undermined public monitoring because the fiscal budget for nuclear energy was a black box with tax-inclusive grants covering the cost of public risk and the disposal of nuclear waste as a by-product of nuclear energy in the name of 'compensation for inconveniences'. The independence of the atomic monitoring agency was seriously weakened, for METI was both the primary regulator and the supporter of nuclear plants. With a strong nuclear village championed by a closed system, risk culture and risk awareness have long been overlooked by overconfident technocrats in METI, and the precautious system of tsunami and system-blackout was never seriously discussed.

The Fukushima event has drastically changed the public image of nuclear energy. An increasing number of Japanese citizens began to question the performance of the central government in regulating the safety of nuclear energy. Owing to that, anti-nuclear civil groups have gained more social support. After Abe's decision in 2014 to restart the nuclear plants, a series of civil protests broke out. The large scale and long consistency of volunteering civil participation really set the milestone for Japan's civil society building, for there have been hardly any civil movements in Japan after the Second World War. However, despite Japanese citizens' growing scepticism regarding the nuclear industry, and the increasing difficulty in reprocessing nuclear waste, Japan still could not abandon its nuclear energy path; the potential possibility of Japan's nuclear rearming was still considered a necessary consideration while Japan's capacity for national defence has been crucial in deterring possible conflicts in East Asia. Owing to the threats, as the only victim of nuclear bombing in human history, Japan's government has been reluctant to relinquish nuclear plants after the Fukushima accident.

Korea shared a very similar economic pattern to that of Japan: authoritarianism-led economic growth with a state-centred nuclear industry. Ku Dowan pointed out that while a massively anti-nuclear government reinforced by democratization in Korea successfully slowed down the path of nuclear proliferation, democracy itself was by no means sufficient in phasing out nuclear use. Developmentalism, a common legacy of East Asian economic models, still dominated the mindset of bureaucracy and the political figures in Korea.

However, as happened in Japan, the Fukushima disaster also awakened Korean public awareness on the social risks of nuclear facilities. With thriving public discussion on nuclear use and alternative choices, the public sphere of energy policies has built on firm ground. Korean citizens thus had more opportunities to recalculate the hidden costs in the current political and economic system, and rethink citizen responsibility to build a more sustainable future.

Like their Korean and Japanese counterparts, Chou's article elaborates on a very similar path of reflexive governance through the mind-awakening of public activism in Taiwan regarding a new landscape of national development strategy. While adopting the same growth-driven economic strategy as Korea, Taiwan also previously strategically introduced highly energy-consuming sectors, such as the petrochemical business, into the country for effective global competition. With the KMT's authoritarian technocracy, nuclear energy emerged as a breakthrough approach for sustaining economic growth, given Taiwan's export-based economic structure. Also, the tight connection between academia and the nuclear industry, along with the black box procedure in the authoritarian regime, also paved the way for policies without citizen monitoring. The nuclear-academic complex blending academic experts and the power industry prevailed in advocating what was presented as an infinite and inexpensive energy supply, making nuclear energy a socially constructed 'necessity' for state development.[1]

However, as Chou describes, the emergence of a robust civil society since the 1990s has removed the veil on this black box. With thriving social activism, the public has realized that great amounts of hidden costs, both social costs and opportunity costs, have been generated by this authoritative energy system. The activism-based three spirals: industrial transition, anti-PM 2.5 and carbon emission, had represented the social robustness of better vision for Taiwan's risk governance. The tri-helix movements directly inspired more social debates and a general sense of commitment. Moreover, those movements did not merely urge a change of national development strategy. Rather, they posed a stringent challenge against the unsatisfactory decision-system supported by technocrats and the state-monopolized power industry. As the three spirals went deeper into the core of Taiwan's developmental state, more and more facts demonstrating that the techno-authoritarianism in Taiwan had failed in governing risks were revealed in front of media and the public. Owing to the momentum, a new era of a more socially responsible, more participatory, and more scientifically robust policy networks were about to take the wheel of state development.

Participation and community power: emancipatory catastrophism

While the authors in the first section demonstrated that East Asia's developmentalism pattern included strong nuclear complexes in nearby countries, the Fukushima catastrophe has awakened the public to the reality that those developmental states are hardly capable of coping with the social risks. The myth of harmless nuclear energy propagated by so-called academic experts, and the promise of state control, no longer elicit unconditional public trust. The civil scepticism against the public policy black box soon led to massive public participation aiming not only to change the given structure through protests and elections, but also to deliver alternative options through citizen collaboration and temperance.

Based on such notions, Yun's chapter describes Korean citizen's autonomous participation in the 'One Less Nuclear Power Plant' project in Seoul to challenge a conservative central regime. This locally organized agenda showed the possibility that the energy issue could be smoothly handled through the enlightened public and a more accountable local city office. Via an on-line opinion survey and community forums, the local office established deliberative energy-governing entities comprising representatives of different civil sectors and public organizations. In addition to widespread grassroots practices in different communities, the city government also launched a series of incentive-based policy measures to facilitate an energy-saving diagnosis and small-scale energy generation. The volunteer works thrived in a wide array of sectors including education, religious and business institutions.

Having said so, the effort in Seoul did not receive as much credit in raising public awareness as it deserved, while most Koreans still considered nuclear power as a necessity because alternatives hardly seemed to be available options. However, under local leadership filled with democratic values, the collaboration between public and private groups at the city level has proved that an alternative energy-governing framework was possible. Although there is a long way to go from Seoul's temporary success, given the fact that the agenda of central government was still dominated by nuclear-favouring technocrats and media, the political experiment in Seoul has shed light on possible better future energy governance.

In Chapter 5, Kao and Chen also present some inspirational autonomous civil actions in Taiwan in leading energy transition to a more diversified and innovative framework. Like other East Asian Tigers, Taiwan shares great similarities with Korea and Japan, for the energy policy in the three countries has long served as a measure of subsidy. Rather than viewing energy as a merchandise or social welfare, inexpensive energy has been considered as the price that government had to offer in state-centred developmentalism. Not only did these states monopolize power generation and distribution, but the energy price, particularly for industrial use, in these three countries was surprisingly low, appearing as the fourth and fifth lowest in a global comparison respectively, in the case of Taiwan and Korea. The skewed and uneven distributed energy structures were aimed to guarantee the continuous investment of the manufacturing sector.

The path-dependent pattern of the developmental state in Taiwan also led the energy transition to take a deviant direction; instead of treating renewable energy as a chance for structural reform, it was viewed as a new strategic business sector. Under this mindset, a series of mega-size renewable energy complexes were built only in pursuit of the profits derived from FIT, while enormous social costs and power inefficiency were neglected in carrying out the policy process. In order to break up the current cumbersome technocratic political structure controlling the energy sector, civil groups in Taiwan tried to exercise their creativity to push for more locally friendly and sustainable energy alternatives. Some of them run 'citizen power plants' by connecting sunroof owners and investors, while others help to build a constant income-generating solar power system for social welfare institutions and rural areas via fundraising. These citizen-based activists collaborated with temples or communities to launch a wide array of local innovations. These examples demonstrated that an energy transition driven by bottom-up initiatives is not only feasible, but also has a great deal of potential regarding Taiwan's future energy supplies.

To solve the historical and structural problems in energy transition necessitates breaking the institutional barrier. In Park's view, the historic dynamic accumulated in the current energy system has handicapped the grassroots movement in its efforts to change the entire energy system (regime) in South Korea. The South Korean Government is addicted to nuclear energy; hence, it hinders the social movement for energy transition and offers insufficient policies for renewable energy. However, the Fukushima disaster, as a kind of ideological shock and catalyst, transformed how people think about the role of energy in their daily lives. By discussing the changing reaction to energy issues, Park offers an overview of how people think about their relations with the environment, society, and science and technology. Park uses the idea of 'energy citizenship', first employed by Devine-Wright, to explain how renewable development could help citizenship mature.

The original energy movement emerged in South Korea in the 1980s. Although South Korea highly depends on nuclear energy to support its energy supply, the nuclear power plants and waste disposal sites were raising a series of debates and conflicts. In these conflicts, at least before 2000, the way the local residents fought against the government was like the struggle of sufferers. Some NGOs had tried to find alternatives, for instance, the decentralized solar power system, to reverse the problematic energy supply system since 2000. Concurrently, the Korean government also introduced renewable-related policies, such as feed-in tariffs, to create a favourable environment for renewable energy development. Nevertheless, the electricity system in South Korea at that time did not prepare for small civil power plants, which discouraged their development. Despite this obstacle, the dynamics of civil power plants by NGOs and locals converged into the 'energy self-sufficient' or 'energy transition' movement.

More radical thinking about nuclear power in South Korea was raised by the Fukushima disaster. In addition, public participation is the core issue of energy transition. As the post-nuclear idea became part of Koreans' perspective, local people voluntarily participated in self-help activities; the vision of phasing out nuclear energy was developed, with some municipal governments creating post-nuclear policies, and citizens involving themselves into an energy-saving and energy-production movement. The 'one less nuclear power plant' movement in Seoul attracted many people. The amount of energy-saving was indeed quite impressive, too. However, regarding the concept of energy citizenship, the most important contribution from these movements to energy transition was the formation of the energy-concerned citizen. The method to facilitate the progress of energy transition and break the institutional barrier Park offers in this book is to increase citizen participation in the whole process.

Energy and social risk: less uncertainty and more procedures

The first two sections in this book had already probed into various problems related to struggle, movement, and participation in energy policy, from the perspectives of governance and participation. The three chapters in the final section are related to different aspects of social risk. However, whether the risk is about radiation exposure, food and health, or the collision between different social interests, their symptoms were all connected to deep structural problems in society: the process of policy formation, the source of expertise and the way to discuss the issues in question, as well as the interaction and communication among government, developers and local residents.

From the STS sociologist's view, Matsumoto and Juraku sharply point out the riskiest issue among all of the Japanese nuclear energy policies: the whole process of policy formation. The siting and disposal of HLW showed that the Japanese government did not enhance the rationality and legitimacy of decision-making on related issues by expanding the sources of knowledge and participation, but rather turned them into issues strengthening public relations problems. METI and NUMO did not accept the assessments by foreign or domestic units like SCJ. In addition, although METI ultimately put forward policy reform programmes, Matsumoto and Juraku criticized the report as delayed or even 're-discovered'. On the one hand, it did not offer many details about what constituted their commitment to reform. On the other hand, it missed the link between the national and international research trends.

The result was a tragic re-discovery of the old. METI's reflection was completely not new because the national practitioners and academic experts in related fields had already elaborated these issues for decades. The reports from the Japanese government or even SCJ were nothing but a correction or rediscovery of these concepts. The whole process of METI's reform was a type II error: undone science. When officials only wanted to manage, while forgetting how the legitimacy of democratic procedures should be carried out, the result was making the process of citizen participation policy into a series of policy display cases. This structural ill-design, resulting in its report and the decision, could not be successful in gaining people's trust or succeed in completing their policy task.

If the government continues to follow and reproduce this problematic procedure, the complexity of the system and the interdependence between units will congregate into many more problems in society, especially when the norms hidden behind the informal groups gradually hollow out the formal rules. When facing problems, technocrats have continued to patch things up again and again. As for those questions, each sector keeps them secret and no one is held responsible.

So the whole characteristics of the disaster formed: no single technical error, no obvious scientific misunderstanding, and no one directly responsible for the catastrophic situation that society faced as a whole. All of them are part of the problem. Technocracy could obtain a political exemption of responsibility, by continuing to respond to these policies, but without solving them. Through opening the social relations on science and technology issues, and their struggle between the structure and policy, Matsumoto and Juraku questioned the whole policy formation process in Japan as a fundamental crisis. In short, before the structure was changed, any report or decision-making produced through this process would be subject to a systematic bias. Without reflection and correction, the disaster is likely to be triggered again and again.

On the other hand, Tateishi's chapter shows that the scientists, the so-called 'experts', could suspend key problems by continually asking questions, rather than taking the initiative to solve the problem. Lacking a good dialogue mechanism, the result was an expert group with only one single source of knowledge, which formed a closed shallow discussion. However, even if the sources of knowledge were diverse, the expert panel might be unfamiliar with basic disciplines and hold insufficient exchanges with civil society experts. This might cause the panel members to take a very roundabout path to reaching a similar conclusion while its decisions were also under highly popular criticism.

This chapter examined the record of the two working groups related to the similar regulative issues. The first one was the Working Group on Risk Management of Low-Dose Radiation Exposure, which is composed entirely of radiation and nuclear engineering experts. They supported the government's opinions and tended to trivialize the uncertainty issue. They seldom discussed this issue; even if they were requested to respond to it, they tended to use terms in their general meaning. By content analysis, Tateshi points out that this group only cited the references from international organizations such as ICRP, but not specific articles or research results to address "specific uncertainty." In this way, this group formed its own "international consensus." Tateishi argues that it impeded an effective dialogue between different parties.

On the other hand, another expert panel was the Working Group of the Food Safety Commission, with diverse knowledge sources. They formed their consensus by concerning themselves with uncertainty. They made a negative self-reference, that is, systemic and reflexive thinking in regard to specific uncertainty, and emphasized their sources of reference. They relied more on specific articles and reports, but less on reports produced by international organizations. Moreover, they criticized these international organizations' reports as difficult to read or to find their sources. Tateishi thought that this process allowed for a deeper discussion in the second group.

In this chapter, it can be said that the landscape, the way to represent the uncertainty in their sources of knowledge, had affected the path of discussion. The difference between the two working groups' compositions had shaped the totally different landscapes on dealing with uncertainty. In their way of speaking, discussing, and debating the uncertainty, their unnoticed attitudes on uncertainty were revealed, and the path to reaching a conclusion tended to become clear. However, not only the composition, but also the ways of meeting and releasing their reports affected the public reaction. In these two cases, both reports resulted in disputes and criticism. Hence, the author reminds us to further reflect on how to achieve a better path for deliberation and communication.

Fan's article confirmed the consequence of the lack of due democratic process and a wide range of communication, namely a series of inevitable conflicts and policy amendments once again. In Taiwan, without a due process, promoting green energy and renewable energy infrastructure, and building nasty polluting factories could actually have shared such a high similarity.

The first issue here is the fairness of interest distribution. From the perspective of the developer, by building wind turbines to promote the development of a village, the residents could benefit from lower electricity prices and subsidies. But from the residents' view, the profit was in fact in the hands of developers. Moreover, in the vision of environmental protection groups looking for a small, decentralized wind power generation system, they argued on the basis of foreign experience: people not only like to own the business, but are also more willing to become involved in the process.

The second issue is risk distribution. From the residents' viewpoint, the health risks and life quality associated with these wind turbines should be questioned. For health risk related to the wind turbine construction, the developer cited a scientific report to prove that the distance between wind turbines and residence does not cause any harm. They believed that before any harm could be proved to be correlated to turbines, they would not change their policy. Although residents could not offer any scientific reports, they used various ways, such as feng shui, common sense, foreign cases, or even the most direct way, physical experience, to explain the depression they felt. As one physician pointed out, their mood also needed to be evaluated as a sort of illness.

However, the developer did not pay attention, and even the local court believed that self-rescue association did not offer precise information to prove that the wind turbines affect their life quality and health. Moreover, the local court also suggested that if the noise does exist, sound insulation windows could be used to improve the situation. This judgment put the power of definition on risk into the hands of the developer, even though the risk was to the residents.

When two groups' values conflict with each other, the government's mission is to clarify the controversy fairly and maintain procedural justice, which in fact it did not do. As Fan argues, what the relevant authorities presented in this case is utilitarian policy thinking: focusing solely on economic interests, and lacking environmental, social, and cultural assessments on sustainability. Procedural justice was ignored and the whole process of risk assessment was a structural disaster, biased by industrial research and its evidence. At the same time, the government's role was so vague that it could not further clarify the scientific controversy.

Why could a science report be so useless in attempting to solve the controversy? Without informed consent, the lack of deliberation, undone science, and other defects in the process the energy-related democracy the government pursued was merely a slogan. A responsible action on defining risk is to settle the dispute, not to ignore it (Renn and Klinke 2016). These are not new problems, but recur again and again, from the development issues in the past to today's energy transition.

Note

1 The power of the nuclear complex almost caused former President Chen to step down in 2003; he decided to discontinue nuclear plant construction.

Reference

Renn, O., and Klinke, A. (2016). Risk perception and its impacts on risk governance. *Oxford Research Encyclopedia of Environmental Science* (May 2016).

Index

Note: Page numbers in *italics* denote references to Figures and Tables

For Product Safety Concerns and Information please contact our EU representative GPSR@taylorandfrancis.com
Taylor & Francis Verlag GmbH, Kaufingerstraße 24, 80331 München, Germany

www.ingramcontent.com/pod-product-compliance
Ingram Content Group UK Ltd.
Pitfield, Milton Keynes, MK11 3LW, UK
UKHW011010180926
475701UK00004B/652
9780367375584